Creative Design of Mechanical Devices

Springer
*Singapore
Berlin
Heidelberg
New York
Barcelona
Budapest
Hong Kong
London
Milan
Paris
Tokyo*

Creative Design of Mechanical Devices

Hong-Sen Yan

Springer

Professor Hong-Sen Yan
Department of Mechanical Engineering
National Cheng Kung University
Tainan 70101
Taiwan
ROC

Library of Congress Cataloging-in-Publication Data

Yan, Hong -Sen, 1951-
Creative design of mechanical devices / Hong-Sen Yan
Includes bibliographical references and index
ISBN 9813083573
1. Machine design.   I. Title
TJ230.Y24   1998
621.8'15--dc21   98-25104
                                                                   CIP

ISBN 981-3083-57-3

This work is subject to copyright. All rights are reserved, whether the whole or part of the material is concerned, specifically the rights of translation, reprinting, reuse of illustrations, recitation, broadcasting, reproduction on micro-films or in any other way, and storage in databanks or in any system now known or to be invented. Permission for use must always be obtained from the publisher in writing.

© Springer-Verlag Singapore Pte. Ltd. 1998
Printed in Singapore

The publisher makes no representation, express or implied, with regard to the accuracy of the information contained in this book and cannot accept any legal responsibility or liability for any errors or omissions that may be made.

Typesetting: Camera-ready by author
SPIN 10689301   5 4 3 2 1 0

PREFACE

This book, based on the author's teaching, research and industrial experiences over the past twenty years, is devoted to presenting engineering creative techniques and a novel creative design methodology for the systematic generation of all possible design configurations of mechanical devices. It provides solid materials to help instructors teach creative design in mechanical engineering. It helps students expand their creative talents effectively. It is also a powerful tool for design engineers to use to come up with fresh concepts to meet new design requirements and constraints, and/or to avoid the patent protection of existing products.

The book is organized in such a way that it can be used for teaching or for self study. Chapter 1 introduces basic concepts of design, design process, and creative design. Chapter 2 explains the formation of mechanical devices, including members, joints, degrees of freedom and topological structures. Chapter 3 deals with engineering creativity, including definitions of creativity, creative process and creative characteristics. Chapter 4 provides rational methods for engineering problem solving, including analysis of existing designs, information search and the checklist method. Chapter 5 introduces creative techniques, including attribute listing, morphological chart analysis and brainstorming. Chapter 6 presents a methodology for the creative design of mechanical devices based on the ideas of generalization and specialization. Chapter 7 deals with the process of generalization, including generalizing principles, generalizing rules and illustrated examples. Chapters 8 and 9 contain algorithms for synthesizing atlases of generalized chains and kinematic chains for mechanical devices, respectively. Chapter 10 provides algorithms of specialization for obtaining all possible topological structures of mechanical devices of interest. Chapters 11, 12, 13, and 14 offer design examples in a step by step sequence to illustrate the applications of the creative design methodology provided in Chapter 6. The exercise problems in each chapter are carefully prepared and organized to help students follow and understand the contents of the text.

The book can be used for undergraduate courses of engineering design or senior design projects. It can also be adopted for graduate courses of advanced machine design, advanced kinematics, or special topics for teaching creative design in mechanical engineering.

The author wishes to express sincere gratitude to his Ph.D. adviser at Purdue University, Dr. Allen S. Hall, Jr., Professor Emeritus, who stimulated the inspiration for this book. The author's association with Dr. Paul Chang, Chairman of Leadwell CNC Manufacturing Co. Ltd. (Taichung, Taiwan), during the past years has been very beneficial to the development of this book. The author wishes also to thank his former graduate students, especially Dr. C. C. Chen, Dr. W. M. Hwang, Dr. C. H. Hsu, Dr. Y. W. Hwang, Dr. L. C. Hsieh, and Dr. F. C. Chen, who contributed much through their theses and dissertations. The author also wishes to thank these ladies, Y. C. Chien, L. S. Liaw, T. Y. Liu, W. L. Chen, and S. H. Hwang, who worked with him over the past ten years on the preparation of the book.

The author believes that this book will fill the needs, both academic teaching and industrial application, for the systematic thinking and generation of new design concepts for mechanical devices. Finally, comments and suggestions for improvement and revision of the book will be very much appreciated.

顏鴻森
Hong-Sen Yan
Department of Mechanical Engineering
National Cheng Kung University
Tainan 70101, Taiwan
February 1998

CONTENTS

Preface v

BACKGROUND TOPICS 1

1 Introduction

    1.1 Design 3
    1.2 Design Process 8
    1.3 Creative Design 9
    1.4 Scope of the Text 13
    1.5 Summary 14
           Problems 14
           References 15

2 Mechanical Devices

    2.1 Mechanical Members 17
    2.2 Joints 21
    2.3 Chains, Mechanisms, and Structures 23
    2.4 Constrained Motion 25
           2.4.1 Planar devices 25
           2.4.2 Spatial devices 27
    2.5 Topological Structures 28
    2.6 Summary 29
           Problems 30
           References 30

CREATIVE PROBLEM SOLVING TECHNIQUES 33

3 Engineering Creativity

    3.1 Definitions 35
    3.2 Creative Process 36
           3.2.1 Preparation phase 36
           3.2.2 Incubation phase 37

		3.2.3 Illumination phase	38
		3.2.4 Execution phase	38
	3.3	Creativity's Characteristics	41
		3.3.1 Creative person	41
		3.3.2 Barriers to creativity	42
		3.3.3 Creative enhancement	45
	3.4	Summary	47
		Problems	47
		References	48

4 Rational Problem Solving

	4.1	Analysis of Existing Designs	49
		4.1.1 Mathematical analysis	50
		4.1.2 Experimental tests and measurements	52
	4.2	Information Search	52
		4.2.1 Literature search	52
		4.2.2 Patent search	54
		4.2.3 File of experts	55
	4.3	Checklist Method	56
		4.3.1 Checklist questions	56
		4.3.2 Checklist transformations	57
	4.4	Summary	62
		Problems	63
		References	63

5 Creative Techniques

	5.1	Introduction	65
	5.2	Attribute Listing	67
		5.2.1 Procedure for attribute listing	67
		5.2.2 Examples	68
	5.3	Morphological Chart Analysis	69
		5.3.1 Characteristics of morphological chart analysis	69
		5.3.2 Procedure for morphological chart analysis	70
		5.3.3 Examples	71
	5.4	Brainstorming	73
		5.4.1 Characteristics of brainstorming	73
		5.4.2 Procedure for brainstorming	74
		5.4.3 Brainstorming group	74
		5.4.4 Brainstorming session	76
		5.4.5 Brainstorming rules	77
		5.4.6 Brainstorming evaluation	78

		5.4.7 Brainstorming report	79
		5.4.8 Examples	79
	5.5	Summary	83
		Problems	83
		References	84

A CREATIVE DESIGN METHODOLOGY 85

6 Creative Design Methodology
  6.1 Introduction ... 87
  6.2 Procedure ... 88
  6.3 Existing Designs ... 89
  6.4 Generalization ... 91
  6.5 Number Synthesis ... 92
  6.6 Specialization ... 92
  6.7 Particularization ... 94
  6.8 Atlas of New Designs ... 94
  Problems ... 95
  References ... 95

7 Generalization
  7.1 Generalized Joints and Links ... 97
  7.2 Generalizing Principles ... 99
  7.3 Generalizing Rules ... 99
  7.4 Generalized (Kinematic) Chains ... 106
  7.5 Examples ... 108
  7.6 Summary ... 115
  Problems ... 115
  References ... 116

8 Generalized Chains
  8.1 Generalized Chains ... 117
  8.2 Link Assortments ... 120
  8.3 Graphs and Chains ... 123
  8.4 Numbers of Generalized Chains ... 126
  8.5 Atlas of Generalized Chains ... 126
  8.6 Summary ... 126
  Problems ... 134
  References ... 134

9 Kinematic Chains

- 9.1. Kinematic Chains 135
- 9.2 Rigid Chains 136
- 9.3 Kinematic Matrices 137
- 9.4 Permutation Groups 140
- 9.5 Enumerating Algorithm 143
  - 9.5.1 Input the numbers of links and degrees of freedom 145
  - 9.5.2 Find link assortments 145
  - 9.5.3 Find contracted link assortments 145
  - 9.5.4 Find incident joint sequences 146
  - 9.5.5 Construct M_{ul} matrices 147
  - 9.5.6 Construct M_{ur} matrices 150
  - 9.5.7 Construct M_{CLA} matrices 152
  - 9.5.8 Transform M_{CLA} matrices to kinematic chains 152
- 9.6 Atlas of Kinematic Chains 153
- 9.7 Summary 157
- Problems 157
- References 158

10 Specialization

- 10.1 Specialized Chains 159
- 10.2 Specializing Algorithm 160
- 10.3 Numbers of Specialized Devices 165
- 10.4 Summary 168
- Problems 169
- References 170

DESIGN PROJECTS 171

11 Clamping Devices

- 11.1 Existing Design 173
- 11.2 Generalization 174
- 11.3 Number Synthesis 175
- 11.4 Specialization 175
- 11.5 Particularization 181
- 11.6 Atlas of New Clamping Devices 181
- 11.7 Remarks 181
- Problems 181
- References 183

12 Motorcross Suspension Mechanisms

- 12.1 Existing Designs 185
- 12.2 Generalization 187
- 12.3 Number Synthesis 188
- 12.4 Specialization 188
- 12.5 Particularization 191
- 12.6 Atlas of New Motorcross Suspension Mechanisms 192
- 12.7 Remarks 192
- Problems 192
- References 195

13 Infinitely Variable Transmissions

- 13.1 Existing Design 197
- 13.2 Generalization 200
- 13.3 Number Synthesis 200
- 13.4 Design Requirements and Constraints 200
- 13.5 Specialization 203
  - 13.5.1 Planetary gear trains with five members 203
  - 13.5.2 Planetary gear trains with six members 204
- 13.6 Particularization 207
- 13.7 Atlas of New Infinitely Variable Transmissions 207
- 13.8 Remarks 207
- Problems 208
- References 209

14 Configurations of Machining Centers

- 14.1 Existing Designs 213
- 14.2 Tree-graph Representations 215
- 14.3 Generalized Tree-Graphs 216
- 14.4 Atlas of Tree-Graphs 217
- 14.5 Specialized Tree-Graphs 218
- 14.6 Atlas of Machining Centers 223
- 14.7 Remarks 228
- Problems 228
- References 228

Appendix   231
Index   237

Background Topics

CHAPTER 1

INTRODUCTION

Design is problem solving. It occurs in everyone's daily life and in every profession. This chapter starts with describing the definitions of design. It proceeds with presenting the processes of engineering design, mechanical design, machine design, and mechanism design. It continues with explaining the nature of creative design, and ends with introducing the application scope of this text.

1.1 Design

Design, a term that derives from the Latin word *designare* meaning to work out, is a creative decision-making process directed toward the fulfillment of human needs. It is the essential purpose of engineering.

According to Webster's New Collegiate Dictionary, *engineering* is defined as the application of science and mathematics by which the properties of matters and the resources of energy in nature are made useful to human beings in structures, machines, products, systems, or processes. Therefore, *engineering design* can be stated as a creative decision-making process to apply the knowledge of science and technology to convert natural resources into devices, products, systems, or processes that are useful to humans.

Engineering design is a creative process that is the essential source of all new devices, products, systems, or processes. It involves considering many different ways to satisfy a need. And most of the time, multiple and even conflicting requirements and constraints must be reconciled. An engineering designer must be able to identify the real needs, to create original ideas, to provide feasible designs for manufacture and maintenance, to consider environmental effects, and to come up with reliable devices, products, systems, or processes with expected performance and with reasonable costs.

Mechanical engineering is a major area of engineering. *Mechanical engineering design*, or *mechanical design* in short, refers to the design of devices, products, systems, or processes of a mechanical nature.

Machine design focuses on the design of machines that consist of movable mechanisms with supporting structures for transmitting motions and forces. A machine should have a certain type of power input, an adequate control system, and an effective work output. *Mechanism design* is concerned mainly with the generation or selection of a particular type of mechanism, the determination of the required numbers and types of members and joints, and the derivations of geometric dimensions of members between joints to achieve the desired constrained motions. Figure 1.1 shows the formation of mechanisms and machines.

Figure 1.2 explains the relations among design, engineering design, mechanical (engineering) design, machine design, and mechanism design. Figure 1.3 shows a vertical machining center. It is the result of an engineering design by a group of engineers with various backgrounds. Figure 1.4 shows the physical design of an automatic tool changing system with a tool change arm. It is the end product of a mechanical design completed by mechanical engineers. Figure 1.5 shows the assembly drawing of a prototype dual-cam tool change device. It is the result of a machine design for exchanging tools between the spindle and the tool magazine. And Figure 1.6 is the schematic representation of the concept of a tool change device to show the relative positions of each member in the mechanism.

In engineering design, all problem solutions are either synthetic or analytical. *Synthesis* is a systematic process, without the iteration procedure, of arranging various elements of concepts in a proper way to generate desired solutions that meet design requirements and constraints. When a problem has a multiplicity of acceptable solutions, it is synthetic. Engineering problems are synthetic in nature, e.g., dimensional synthesis (rigid body guidance, coordination of input-output motion, function generation, and path generation) of four-bar linkages. *Analysis* is the process for verifying an existing solution. When a problem has a single right answer it is analytical in nature. Analysis is by far the most common form of engineering problem solving. Many activities of engineering tasks are of this kind, e.g., kinematic (position, velocity, acceleration) analysis of four-bar linkages.

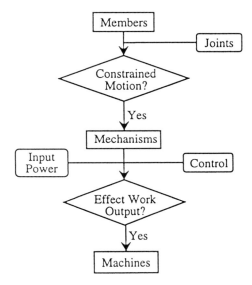

Figure 1.1 Formation of mechanisms and machines

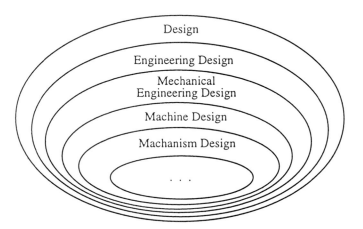

Figure 1.2 Types of design

Figure 1.3 Machining center – an engineering design (see Appendix)
(Courtesy of Leadwell Company)

Figure 1.4 Automatic tool changing system – a mechanical design
(Courtesy of Leadwell Company)

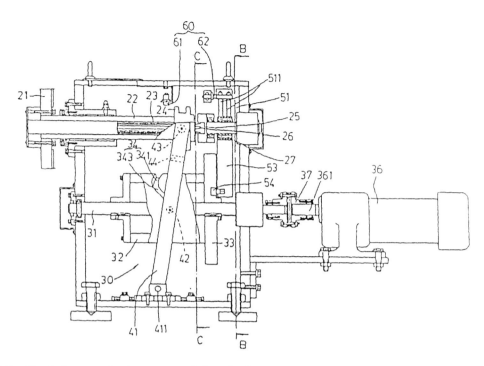

Figure 1.5 Tool change device – a machine design
(U.S. patent No. 5,129,140)

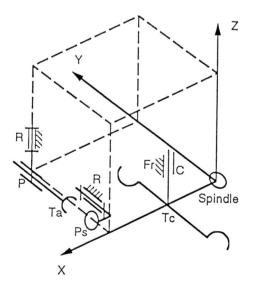

Figure 1.6 Tool change mechanism – a mechanism design

For most engineering problems, direct solutions based on synthesis are not normally available. In such a case, an existing design or a tentative design should be identified or proposed first. Then, an iterative procedure, based on the techniques of numerical analysis and optimization, should be carried out to reach an acceptable solution. Figure 1.7 shows the relations among design, synthesis, and analysis for engineering solutions.

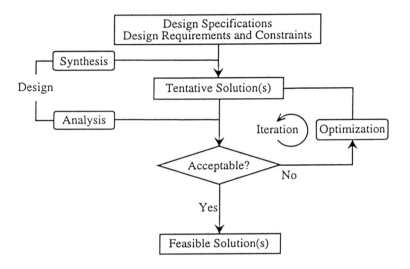

Figure 1.7   Design, synthesis, and analysis for engineering solutions

1.2 Design Process

Engineering design creates useful devices, products, systems, or processes. In making out these results, a logical sequence of events should be followed to ensure the success of the devices, products, systems, or processes. This logical sequence of events is known as the *design process*.

Traditionally, design is mostly learned by experience, and a unified design process is difficult to identify. However, experienced engineers can indicate that there are always some general events that are common in designing various devices, products, systems, or processes. Therefore, a design process, although it may not be unified, should be identifiable and can serve as a useful reference in executing different engineering projects for designers.

Various descriptions of design process for various types of design are available in various literatures. However, for each type of design, most of these descriptions tend to be similar. Figure 1.8 shows a typical engineering design process that consists of the following steps: recognition of the need, definition of the problem, creation of a design, preparation of a model, evaluation of the model, and communication of the design. Figure 1.9 shows a typical process of mechanical (engineering) design, which starts with problem definition and ends with manufacture drawings to communicate the design. And Figure 1.10 shows the process of designing mechanisms and machines.

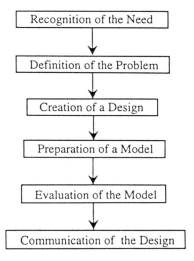

Figure 1.8 Engineering design process

In designing a device, product, system, or process, feedback among design steps in the various design processes mentioned above is usually needed to improve the design.

1.3 Creative Design

Creation is the process of bringing something new into existence.

In the engineering design step of "creation of a design", designers need to make adequate judgments and generate feasible concepts as basic design configurations for the step of "preparation of a model".

A *concept* is an abstract idea and is a basic way of performing a required task for solving a specific design problem. *Conceptual design* involves the process of obtaining a pool of feasible concepts from which the most promising one is selected. This is the process of creation, the most difficult and least understood step in the design process.

In the long history of human civilization, many ingenious designs have been invented. However, when one attempts to trace the birth of these designs, one usually falls into the mysterious world of invention without clear answers. Figure 1.11 shows a water container that was used around 4,000 BC in China. By placing this device on a water surface, it automatically fills with water according to the concepts of buoyancy force and center of gravity - which were not known in that period. While technology has leapt forward, the progress of creative thinking has barely increased. So far, no existing method is available to guide designers directly and precisely to invent devices, products, systems, or processes.

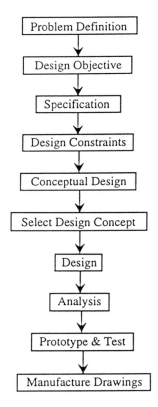

Figure 1.9 Mechanical design process

1.3. Creative Design ▪ 11

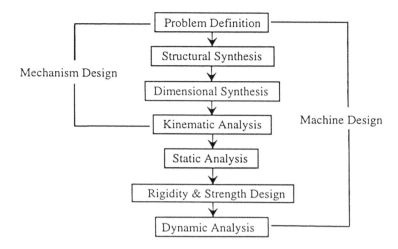

Figure 1.10 Mechanism and machine design process

Figure 1.11 An ancient Chinese water container

Traditionally, design concepts are obtained based on the designers' knowledge, experience, imagination, ingenuity, inspiration, and/or intuition. Figure 1.12 shows a design named "endless screw" invented by Leonardo da Vinci in the fifteenth century, and this is supposed to be the original concept for today's "roller gear cams" (Figure 1.13) that are widely used in the machine tool industry.

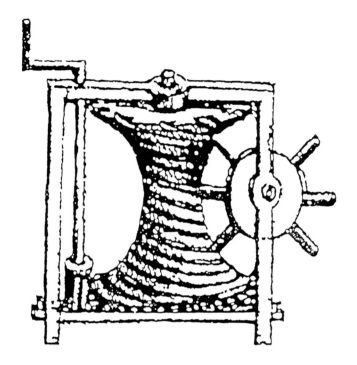

Figure 1.12   Leonard da Vinci's endless screw

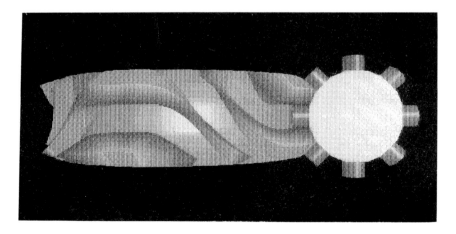

Figure 1.13   A roller gear cam (see Appendix)

1.4 Scope of the Text

Design concepts may be brought into existence with the aid of a stimulant technique serving as a catalyst for the concept derivation. Many problem solving techniques, such as analysis of existing designs, information search, checklist method, attribute listing, morphological chart analysis, brainstorming, ... etc., are available for helping idea generation in conceptual design. Chapters 3-5 are intended to present these techniques for creative problem solving. These organized methods, nevertheless, are not quite systematic, and usually cannot be guaranteed to come up with desired solutions.

Experience has shown that an engineering designer spends more then the time allowed to arrive at the first design concept. If the process of creative design can be better organized, made systematic and even automatic, more innovative designs may be invented.

The creation of basic design configurations of mechanical devices can be generally classified into two categories. The creation of some or all possible design concepts that have not existed before and the creation of more or all possible design concepts that have the same or required topological characteristics as existing designs, Figure 1.14. To come up with something totally new is an invention, is an enjoyable experience, and is probably the most difficult and exciting phase in the design process. However, most cases of designing mechanical devices are not to invent totally new designs, but to modify existing ones to meet new design requirements and constraints, to avoid patent protections or for cross licensing. Clients and customers usually require improvements rather than novelties. Therefore, making variations on established subjects is an important feature of design activities. It is also the way in which much creative thinking actually develops.

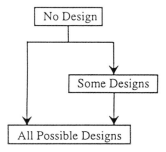

Figure 1.14 **Types of creative design of mechanical devices**

Chapter 2 explains what are mechanical devices. Chapter 6 introduces a creative design methodology, based on available existing designs, for the systematic and precise discovering of all possible alternative design concepts of mechanical devices that perform the same or required tasks as existing designs. Chapters 7-10 describe the major steps of this design methodology in detail. And Chapters 11-14 provide illustrated design examples.

1.5 Summary

Creation is the process of bringing something new into being. Design is a creative decision-making process directed towards the fulfillment of human needs. Engineering design is a process to convert natural resources into engineering products by applying knowledge of science and technology. Mechanical design refers to the design of products of a mechanical nature. Machine design focuses on the design of machines for transmitting motions and forces. Mechanism design concerns mainly the topological structure and the geometric dimensions of the design to achieve constrained motion.

Engineering problems are synthetic in nature and direct solutions based on synthesis are not normally available.

A design process is a logical sequence of events to ensure the success of designing devices, products, systems, or processes. And feedback among design steps in various design processes is usually needed to improve the design.

A concept is an abstract idea of performing a required task for solving a specific design problem. Conceptual design is the process of creation, and is the most difficult and least understood step in the design process.

The purpose of this text is to present engineering creative techniques and a creative design methodology for the systematic and precise generation of all possible design configurations of mechanical devices.

Problems

1.1 Identify at least three different types of design in various areas other than engineering.
1.2 Describe at least three definitions of design in various areas other than engineering.
1.3 Explain the differences between engineering design and other types of design.

1.4 Identify an engineering design product from the home appliances industry and describe the characteristics of this design.
1.5 Identify a machine design product from the automotive industry and describe the characteristics of this design.
1.6 Provide three examples of design that are synthetic and analytic, respectively, in nature.
1.7 What is the difference between the process of designing a bicycle and the process of painting a picture?
1.8 Name five significant mechanical products in the twentieth century.
1.9 Name three consumer products of the past twenty years that you regard as most creative.

References

Shoup, T. E., Fletcher, L. S., and Mochel, E. V., Introduction to Engineering Design, Prentice-Hall, 1981.

Webster's New Collegiate Dictionary, G. & C. Merriam, 1981.

Yan, H. S., "What is design?," Proceedings of the 9th National Conference of the Chinese Society of Mechanical Engineers, Kaoshung, Taiwan, November 27-28, 1992, pp.1-5.

Yan, H. S., Mechanisms, Tung Hua Books, 1997.

Yan, H. S., Yiou, C. W., and Chao, P. C., "Cutter exchanging apparatus incorporated in a machine," U.S. Patent No. 5,129,140, July 14, 1992.

CHAPTER 2

MECHANICAL DEVICES

A *mechanical device* is a piece of equipment, mechanical in nature, designed to serve a special purpose or perform a special function. It generally consists of mechanical members connected by joints. These members are so formed and connected that they transmit constrained motions by moving upon each other as *mechanisms*, or they transmit forces without any relative motion as *structures*. Four-bar linkages, used in various applications for transmitting motions, are typical examples of mechanisms. Aircraft landing gears, for the purpose of absorbing impact forces during landing, are good examples of structures. The types and numbers of mechanical members and joints, and the incidences between them, characterize the *topological structure* of mechanical devices.

This chapter presents mechanical members and joints, the definitions of chains, mechanisms and structures, the concept of constrained motion for both planar and spatial devices, and the identification of the topological structure of mechanical devices. Figure 2.1 shows the formation of mechanical devices.

2.1 Mechanical Members

Mechanical members are resistant bodies that collectively form mechanical devices. They can be rigid members (as kinematic links, sliders, rollers, gears, cams, power screws, shafts, keys, rivets, ... etc.), flexible members (as springs), or compression members (as airs or fluids). Compression members and those for the purpose of fastening two or more members together are not of interest here. Only those members whose function is to provide possible relative motion with others are presented.

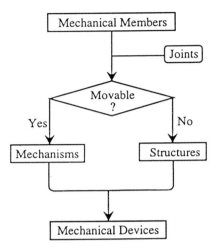

Figure 2.1 Formation of mechanical devices

There are numerous types of mechanical members. In what follows, the functional descriptions and schematic representations of basic mechanical members are described.

Kinematic link

A *kinematic link* (K_L), or just a *link* in short, is a rigid member for holding its joints apart and transmitting motions and forces. Globally speaking, any rigid mechanical member is a kinematic link. Links can be classified based on the number of incident joints. A *separated link* is one with zero incident joints. A *singular link* is one with one incident joint. A *binary link* is one with two incident joints. A *ternary link* is one with three incident joints. A *quaternary link* is one with four incident links. An L_i-link is one with i incident joints. Graphically, a link with i incident joints is symbolized by a crosshatched, i-sided polygon with small circles on the vertices indicating incident joints. Figure 2.2(a) shows the schematic representations of a separated link, singular link, binary link, and ternary link.

Slider

A *slider* (K_P) is a link that has either rectilinear or curvilinear translation. Its purpose is to provide a sliding contact with an adjacent member. Figure 2.2(b) shows the schematic representations of a rectilinear slider and a curvilinear slider.

Roller

A *roller* (K_O) is a link for the purpose of providing rolling contact with an adjacent member. Figure 2.2(c) shows the schematic representation of a simple roller.

Gear

Gears (K_G) are links that are used, by means of successively engaging teeth, to provide positive motion from a rotating shaft to another that rotates, or from a rotating shaft to a body that translates. Gears can be classified as spur gears, bevel gears, helical gears, and worm and worm gears. Figure 2.2(d) shows the schematic representation of a typical gear.

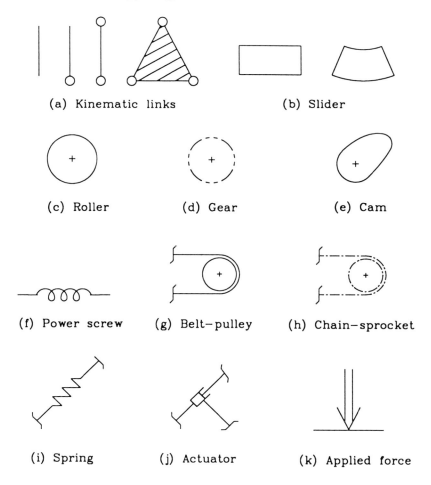

Figure 2.2 Schematic representations of mechanical members

Cam

A *cam* (K_A) is an irregularly shaped link that serves as a driving member and imparts a prescribed motion to a driven link called *follower*. Cams can be classified as wedge cams, disk cams, cylindrical cams, barrel cams, conical cams, spherical cams, roller gear cams, and others. Figure 2.2(e) shows the schematic representation of a disk cam.

Power screw

Power screws (K_H) are used for transmitting motions in a smooth and uniform manner. They may also be thought of as linear actuators that transform rotary motion into linear motion. Figure 2.2(f) shows the schematic representation of a power screw.

Belt

Belts (K_B) are tension members for power transmissions and conveyers. They obtain their flexibility from distortion of the material, and motion is usually transmitted by means of friction between belts and the corresponding *pulleys* (K_U). Belts can be classified as flat belts, V belts, and timing belts. Figure 2.2(g) shows the schematic representation of a belt with its pulley.

Chain

Chains (K_C) are also tension members for power transmissions and conveyers. They are made from small rigid parts that are joined in such a manner as to permit relative motion of the parts, and motion is usually transmitted by positive means, i.e., *sprockets* (K_K). Chains can be classified as hosting chains, conveying chains, and power transmission chains. Figure 2.2(h) shows the schematic representation of a roller chain with its sprocket.

Spring

Springs (K_S) are flexible members. They are used for storing energy, applying forces, and making resilient connections. Springs can be classified as wire springs, flat springs, and special-shaped springs. Figure 2.2(i) shows the schematic representation of a typical spring.

Actuator/shock absorber

An *actuator/shock absorber* (K_T) consists of a *piston* (K_I) and a *cylinder* (K_Y) with a kind of compression member bounded by the piston and the cylinder. Its purpose is to provide a damping action between the members adjacent to the actuator. Figure 2.2(j) shows the schematic representation of an actuator.

Applied force

An *applied force* (K_D) is an external force acting on a mechanical device, especially a clamping device, to provide the necessary clamping force. Figure 2.2(k) shows the schematic representation of an applied force.

2.2 Joints

In order to make mechanical members useful, they must be connected by certain means. That part of a mechanical member that is connected to a part of another member is called an *element*. Two elements that belong to two different members and are connected together form a *kinematic pair* or *joint*.

In what follows, the functional descriptions and schematic representations of basic kinematic pairs are introduced.

Revolute pair/turning pair

For a *revolute* or *turning pair* (J_R), the relative motion between two incident members is rotation about an axis. Figure 2.3(a) shows its schematic representation.

Prismatic pair/sliding pair

For a *prismatic* or *sliding pair* (J_P), the relative motion between two incident members is translation along an axis. Figure 2.3(b) shows its schematic representation.

Rolling pair

For a *rolling pair* (J_O), the relative motion between two incident members is pure rolling without slipping. Figure 2.3(c) shows its schematic representation.

Gear pair

For a *gear pair* (J_G), the relative motion between two incident members is the combination of rolling and sliding. Figure 2.3(d) shows its schematic representation.

Cam pair

For a *cam pair* (J_A), the relative motion between two incident members is the combination of rolling and sliding. Figure 2.3(e) shows its schematic representation.

Wrapping joint

For a *wrapping joint* (J_W), there is no relative motion between two incident members. However, one of the members (pulley or sprocket) rotates about its center. Figure 2.3(f) shows its schematic representation.

Helical pair/screw pair

For a *helical or screw pair* (J_H), the relative motion between two incident members is helical. Figure 2.3(g) shows its schematic representation.

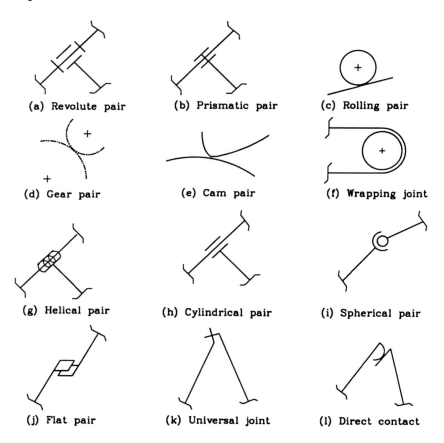

Figure 2.3 Schematic representations of joints

Cylindrical pair

For a *cylindrical pair* (J_C), the relative motion between two incident members is a combination of a rotation about an axis and a translation parallel to the same axis. Figure 2.3(h) shows its schematic representation.

Spherical pair

For a *spherical pair* (J_S), the relative motion between two incident members is spherical. Figure 2.3(i) shows its schematic representation.

Flat pair

For a *flat pair* (J_F), the relative motion between two incident members is planar. Figure 2.3(j) shows its schematic representation.

Universal joint

For an *universal joint* (J_U), the relative motion between two incident members is spherical. Figure 2.3(k) shows its schematic representation.

Direct contact

In structures, especially clamping devices, two members are sometimes connected by a *direct contact* (J_D) due to an applied force, and there should be no relative motion between two incident members. Figure 2.3(l) shows its schematic representation.

2.3 Chains, Mechanisms, and Structures

When several links are connected together by joints, they are said to form a *link-chain* or just a *chain* in short. An (N_L, N_J) chain is a chain with N_L links and N_J joints. Graphically, a joint is symbolized by a small circle, and a link with i incident joints is symbolized by an i-sided, crosshatched polygon with small circles as vertices. For the sake of simplicity, a joint symbolized by a small circle is referred as a revolute joint in general.

A *walk* of a chain is an alternating sequence of links and joints beginning and ending with links, in which each joints is incident with the two links immediately preceding and following it. A *path* of a chain is a walk in which all the links are distinct. If any two links of a chain can be joined by a path, the chain is said to be *connected*; otherwise the chain is *disconnected*. Figure 2.4(a) shows a (5,4) disconnected chain with a separated link (link 5), and Figure 2.4(b) shows a (5,5) connected chain with a singular link (link 5). If every link in the chain is connected to at least two other links, the chain forms one or several closed loops and is called a *closed chain*. A connected chain that is not closed is an *open chain*. A *bridge-link* in a chain is a link whose removal results in a disconnected chain. Figure 2.4(c) shows a (7,8) closed chain with a bridge-link (link 1). The connected chain shown in Figure 2.4(b) is also an open chain.

A *kinematic chain* generally refers to a movable chain that is connected, closed, without any bridge-link, and with revolute pairs only. If one of the links in a kinematic chain is fixed as the *ground link* (K_F), it is a *mechanism*. Figure 2.5(a) shows a (6,7) kinematic chain, and Figure 2.5(b) shows its corresponding mechanism obtained by grounding link 1 in the chain.

A *rigid chain* refers to an immovable chain that is connected, closed, without any bridge-link, and with revolute pairs and simple joints only. If one of the links in a rigid chain is fixed or grounded, it is a *structure*. Figure 2.6(a) shows a (4,5) rigid chain, and Figure 2.6(b) shows its corresponding structure by grounding link 1 in the chain.

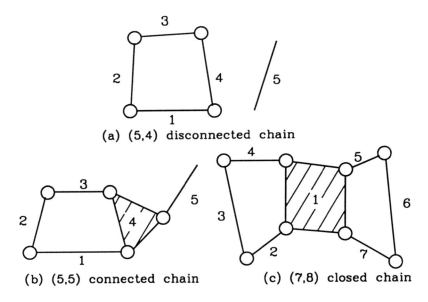

(a) (5,4) disconnected chain

(b) (5,5) connected chain

(c) (7,8) closed chain

Figure 2.4 Types of (link-)chains

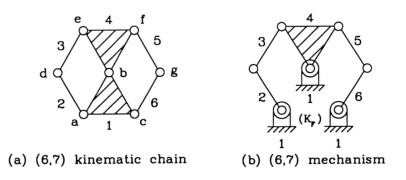

(a) (6,7) kinematic chain

(b) (6,7) mechanism

Figure 2.5 A (6,7) kinematic chain and mechanism

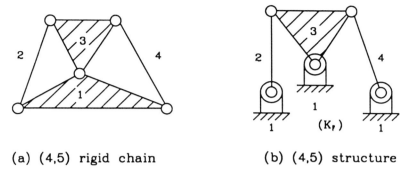

(a) (4,5) rigid chain (b) (4,5) structure

Figure 2.6 A (4,5) rigid chain and structure

2.4 Constrained Motion

The number of *degrees of freedom* (*F*) of a mechanical device determines how many independent inputs the device must have in order to fulfill a useful engineering purpose. In another words, it is the number of independent coordinates needed to specify the relative positions of members in the mechanical device.

A mechanical device with a positive number of degrees of freedom and with the same number of independent inputs is a *mechanism* and has constrained motion. *Constrained motion* means that when any point on an input member of the device is moved in a prescribed way, all other moving points of the device have uniquely determined motions. If the number of the independent inputs is less than the number of degrees of freedom, the mechanical device is generally unconstrained. A mechanical device with zero degrees of freedom is a *structure* that is overconstrained and immovable. If a mechanical device possesses negative degrees of freedom, it is a structure with redundant constraints.

2.4.1 Planar Devices

For planar mechanical devices, a member has three degrees of freedom consisting of translational motions along two mutually perpendicular axes and a rotational motion about any point. The number of *degrees of freedom*, F_p, of a planar mechanical device with N_L members and N_{Ji} joints of type i is:

$$F_p = 3(N_L-1) - \Sigma N_{Ji} C_{pi} \qquad (2.1)$$

where C_{pi} is the number of *degrees of constraint* of i-type joint.

[Example 2.1]
Calculate the number of degrees of freedom for the horizontal tail control mechanism, with two independent inputs, of an aircraft shown in Figure 2.7. Member 2 is an input (I) from the control stick, the actuator (members 8 and 9) is another input (II) for the purpose of stability augmentation, and member 7 is the output link.

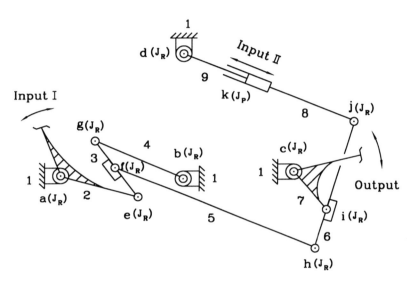

Figure 2.7 An aircraft horizontal tail control mechanism with two inputs

This is a planar mechanism with nine members (links 1, 2, 3, 4, 5, 6, 7, 8, and 9) and eleven joints consisting of ten revolute joints ($a, b, c, d, e, f, g, h, i,$ and j) and one prismatic joint (k). Therefore, $N_L=9$, $C_{pR}=2$, $N_{JR}=10$, $C_{pP}=2$, and $N_{JP}=1$. Based on Equation 2.1, the number of degrees of freedom, F_p, of this device is:

$$F_p = 3(N_L-1) - (N_{JR}C_{pR}+N_{JP}C_{pP})$$
$$= (3)(9-1) - [(10)(2)+(1)(2)]$$
$$= 2$$

Therefore, the motion of this device is constrained.

If the stability augmented system, i.e., input II, is not activated, links 8 and 9 have no relative motion with each other. It then becomes a mechanism with eight members and ten revolute joints, and its number of degrees of freedom, F_p, is:

$$F_p = 3(N_L-1) - N_{JR}C_{pR}$$
$$= (3)(8-1) - (10)(2)$$

$$= 1$$

Therefore, the motion of this device is still constrained.

2.4.2 Spatial Devices

For spatial mechanical devices, a member has six degrees of freedom consisting of translational motions along three mutually perpendicular axes and three rotational motions about these axes. The number of *degrees of freedom*, F_s, of a spatial mechanical device with N_L members and N_{Ji} joints of type i is:

$$F_s = 6(N_L-1) - \Sigma N_{Ji}C_{si} \qquad (2.2)$$

where C_{si} is the number of *degrees of constraint* of i-type joint.

[Example 2.2]
Explain if the Macpherson strut suspension mechanism shown in Figure 2.8 has constrained motion. The input of this device is from the wheel to member 3.

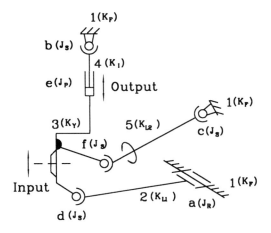

Figure 2.8 Macpherson strut suspension mechanism

This is a spatial mechanism with five members (K_F, link 1; K_{L1}, link 2; K_Y, link 3; K_I, link 4; K_{L2}, link 5) and six joints consisting of one revolute joint (a), one prismatic joint (e), and four spherical joints (b, c, d, and f). Therefore, $N_L=5$, $C_{sR}=5$, $N_{JR}=1$, $C_{sP}=5$, $N_{JP}=1$, $C_{sS}=3$, and $N_{JS}=4$. Based on Equation 2.2, the number of degrees of freedom, F_s, of this device is:

$$F_s = 6(N_L-1) - (N_{JR}C_{sR}+N_{JP}C_{sP}+N_{JS}C_{sS})$$
$$= (6)(5-1) - [(1)(5)+(1)(5)+(4)(3)]$$
$$= 2$$

This is still a useful device, since the rotation of link 5 about the axis through the centers of spherical joints c and f is an extra degree of freedom that does not affect the input-output relation of the system.

2.5 Topological Structures

Two chains or mechanical devices are said to be *isomorphic* if they have the same topological structures. The *topological structure* of a chain, mechanism, structure, or mechanical device is characterized by its types and numbers of links and joints, and the incidences between them. The *isomorphism* of mechanical devices can be identified based on the concept of topology matrix.

The *topology matrix*, M_T, of an (N_L,N_J) mechanical device is an N_L by N_L matrix. Its diagonal element $e_{ii}=u$ if the type of member i is u; its upper off-diagonal entry $e_{ik}=v$ ($i<k$) if the type of the joint incident to members i and k is v, its lower off-diagonal entry $e_{ki}=w$ if the assigned name of the joint is w, and $e_{ik}=e_{ki}=0$ if members i and k are not adjacent.

[Example 2.3]
Identify the topological structure of the cam-roller-actuator mechanism shown in Figure 2.9.

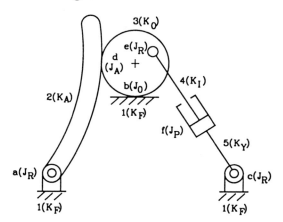

Figure 2.9   A cam-roller-actuator mechanism

This mechanism has five members and six joints. K_F (member 1) is the ground link, K_A (member 2) is a cam, K_O (member 3) is a roller, K_I (member 4) is a piston, and K_Y (member 5) is a cylinder. The joint (a) incident to K_F and K_A is a revolute pair (J_R); the joint (b) incident to K_F and K_O is a rolling pair (J_O); the joint (c) incident to K_F and K_Y is a revolute pair (J_R); the joint (d) incident to K_A and K_O is a cam pair (J_A); the joint (e) incident to K_O and K_I is a revolute pair (J_R); and the joint (f) incident to K_I and K_Y is a prismatic pair (J_P). The topology matrix, M_T, of this mechanical device is:

$$M_T = \begin{bmatrix} K_F & J_R & J_O & 0 & J_R \\ a & K_A & J_A & 0 & 0 \\ b & d & K_O & J_R & 0 \\ 0 & 0 & e & K_I & J_P \\ c & 0 & 0 & f & K_Y \end{bmatrix}$$

[Example 2.4]
Identify the topological structure of the Macpherson strut suspension mechanism shown in Figure 2.8.

This device also has five members and six joints. Member 1 (K_F) is the ground, member 2 (K_{L1}) is a kinematic link, member 3 (K_Y) is the wheel link and the cylinder of the shock absorber, member 4 (K_I) is the piston of the shock absorber, and member 5 (K_{L2}) is another kinematic link. The joint (a) incident to K_F and K_{L1} is a revolute pair (J_R); the joint (b) incident to K_F and K_I, and the joint (c) incident to K_F and K_{L2} are spherical pairs (J_s); the joint (d) incident to K_{L1} and K_Y is also a spherical pair (J_s); the joint (e) incident to K_Y and K_I is a prismatic pair (J_P); and the joint (f) incident to K_Y and K_{L2} is another spherical pair (J_s). The topology matrix, M_T, of this device is:

$$M_T = \begin{bmatrix} K_F & J_R & 0 & J_S & J_S \\ a & K_{L1} & J_S & 0 & 0 \\ 0 & d & K_Y & J_P & J_S \\ b & 0 & e & K_I & 0 \\ c & 0 & f & 0 & K_{L2} \end{bmatrix}$$

2.6 Summary

A mechanical device consists of mechanical members connected by joints. Mechanical members are resistant bodies for transmitting motions and forces. To make mechanical members useful, they must be connected by joints.

When several links are connected together by joints, they are said to form a chain, which can be connected or disconnected, closed or open. A kinematic chain refers to a movable chain that is connected, closed, without any bridge-link, and with revolute pairs only. A rigid chain is one that is not movable.

The number of degrees of freedom of a mechanical device determines how many independent inputs the device must have in order to fulfill a useful engineering purpose. A mechanical device with a positive number of degrees of freedom and with the same number of the independent inputs is a mechanism, and a mechanical device with zero or negative degrees of freedom is a structure.

The types and numbers of mechanical members and joints, and the incidences between them characterize the topological structure of mechanical devices. The concept and definition of the topology matrix is a neat and powerful tool for testing isomorphism of generated mechanical devices in the process of creative design.

Problems

2.1 List the various mechanical members in a bicycle.
2.2 Name at least five applications of revolute joints in your house.
2.3 Name three applications of spherical joints.
2.4 Identify various members and joints in a mechanical lock.
2.5 Calculate the number of degrees of freedom for the cam-roller-actuator mechanism shown in Figure 2.9.
2.6 Identify the topology matrix for the aircraft horizontal tail control mechanism shown in Figure 2.7.
2.7 Name one planar mechanism with six members, describe its function, sketch its schematic diagram, identify its topology matrix, and calculate its number of degrees of freedom.
2.8 Name one spatial mechanism with at least three members, describe its function, sketch its schematic diagram, identify its topology matrix, and calculate its number of degrees of freedom.
2.9 Name one application of structures, describe its function, sketch its schematic diagram, identify its topology matrix, and calculate its number of degrees of freedom.

References

Harary, F. and Yan, H. S., "Logical foundations of kinematic chains: graphs, line graphs, and hypergraphs," ASME Transactions, *Journal of Mechanical Design*, Vol. 112, No. 1, 1990, pp. 79-83.

Hwang, W. M. and Yan, H. S., "Atlas of basic rigid chains," Proceedings of the 9th Applied Mechanisms Conference, Session IV-B, No. 1, Kansas City, Missouri, October 28-30, 1985.

Hwang, Y. W., An Expert System for Creative Mechanism Design, Ph.D. dissertation, Department of Mechanical Engineering, National Cheng Kung University, Tainan, Taiwan, May 1990.

Yan, H. S., Mechanisms, Tung Hua Books, 1997.

Creative Problem Solving Techniques

CHAPTER 3

ENGINEERING CREATIVITY

The history of civilization is a history of humankind's creative efforts through the centuries. Engineering is a creative profession, and the teaching of creativity has become a necessity in engineering education. Few people can hope to reach the creative genius of a da Vinci or Edison. However, everyone is born creative. And one can learn to develop and use more efficiently his creative talent. For this purpose, this chapter clears up definitions relating to creativity, describes the creative process, and explains creativity's characteristics.

3.1 Definitions

Creativity is the mental ability to order, structure, pattern, arrange or compose in new ways, a chosen set of elements according to principles of economy and aesthetics. It is a defining of previously unknown things.

In science and engineering work, the term creativity is sometimes used interchangeably with innovation. However, these two terms are not synonymous. They are actually complementary to each other in the design process. *Innovation* is the introduction of a new idea, method, or device. It includes the recombination and/or modification of existing ideas, methods, or devices. *Creativity* is the origination of a concept to fulfill a human need. And innovation may or may not respond to a human need, and may or may not be valuable. In effect, creativity is innovation to meet a need.

Invention is a process to produce something useful for the first time through the use of imagination, thinking, studying, experiment, and/or experience. From the engineering point of view, a creation is an invention that can be developed for use by others. Therefore, an invention is the result of creative thought.

3.2 Creative Process

In generating fresh ideas, new methods or novel devices, it is believed that there exists a logical step by step procedure, namely the *creative process*. It may include preparation phase, incubation phase, illumination phase, and execution phase. The major steps for each phase involved in the process are shown in Figure 3.1.

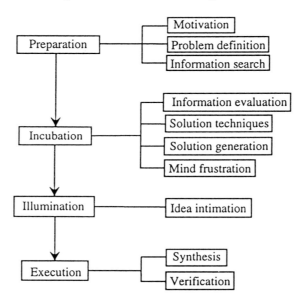

Figure 3.1 Creative process

3.2.1 Preparation Phase

The creative process begins with the recognition of a need for the solution of a problem and the preparation of information for the defined problem.

Motivation

A creative effort requires intense thinking and usually consumes considerable energy. Strong individual desire is then needed to generate the determination to supply and expend the required energy.

Such highly concentrated desires are often maintained based on motivations. And motivations usually differ from individual to individual. It may be natural curiosity, personal values, family pressure, career interests, or it may be job related.

Problem definition

Once an engineer is motivated to embark on a design project, he must define the problem precisely in order to orient his thinking toward the problem's solution. Specifications, including design requirements and constraints, of the problem must be established. However, the problem is usually not totally understood in this stage of the design process. Experienced engineers consider all aspects of the problem as they prepare for a solution. This preparation is essential to the process of obtaining a feasible and desired solution.

Information search

In this step, the engineer should gather as much information as possible about the problem. This can be achieved by literature search, site visit, and/or expert consultation. In general, engineers usually conduct a literature search to learn what others have done about related problems. They may study textbooks, research dissertations, journal articles, technical reports, patents, and commercial catalogs. Visits of some sites relevant to the problem and a thorough discussion with persons familiar with various aspects of the problem are sometimes valuable for gathering information. The content of the information search will be described in more detail in section 4.2.

3.2.2 Incubation Phase

Once the problem has been defined and sufficient information has been obtained, the next stage of the creative process is incubation. This phase is critical to creative solutions. It is usually a long and lonely process, and it is a period for allowing one's subconscious to work. The incubation phase may include steps of information evaluation, solution techniques, solution generation, and mind frustration.

Information evaluation

The initial step of the incubation phase is to evaluate the information obtained in the preparation phase. This may involve the classification of available data or the rearrangement of the original information into other forms. This step allows the engineer to understand thoroughly the available information.

Solution techniques

In addition to evaluating the basic information, the engineer should consider the various possible techniques for solving the problem. The solution techniques may involve experimental techniques, analytical methods, numerical iterations, or graphical means. For most engineering problems, the engineer uses several of these techniques in seeking an optimal solution.

Solution generation

Based on the knowledge gained through the earlier steps of the creative process, some solutions to the problem may seem possible. The engineer explores in great detail all these possible solutions and various combinations that may lead to a satisfactory answer.

Mind frustration

In trying to develop a solution, one is often faced with a good deal of frustration. Most of the time, no solution may seem appropriate, and no amount of thinking about the alternatives may seem to produce a fruitful idea. The nagging of the mind may create tension, anxiety, or emotional stress. A certain amount of discomfort often can act to stimulate the creative process. The problem is quite often solved at this stage. If the problem is not solved to the satisfaction of the engineer, at least he will become familiar with the task down to the most minute detail.

3.2.3 Illumination Phase

The phase after incubation of the creative process is illumination.

When the creative idea occurs in a flash, usually during a period of rest or during engagement in an activity completely foreign to the problem, illumination has taken place. It is the point at which one's conscious mind receives a hint of a suitable solution.

Most people at one time or another have experienced illumination, the sudden, spontaneous appearance of an answer to a problem at unpredictable times when the mind is seemingly engaged in other matters. Experts indicate that one's subconscious works harder in probing deeper for ideas when the conscious mind is at rest.

The phase of illumination is not yet clearly understood. Its success depends upon the thoroughness of the work in the preceding steps.

3.2.4 Execution Phase

The final phase of the creative process is execution, which includes synthesis and verification.

Synthesis

Synthesis is the composition or combination of parts or elements so as to form a whole. It is that step in the creative process during which the engineer uses his knowledge to link the diverse parts of the solution to form a final solution.

Verification

Once the ideas have been synthesized to come up with an organized solution to the problem, the next step is to verify the solution. When many ideas have been generated, they must be evaluated to narrow down the number of alternatives to the few that offer the greatest potential benefits. Judgment requires evidence to prove if an idea is actually worthwhile and this can be found through analysis, experimentation, or sometimes the opinion of experts.

It should be noted that not all of the steps for each phase of the creative process occur in the solution to every problem. The nature of the problem and the solution required will dictate which of the steps could be omitted and which will overlap. Creative ideas may also be generated by applying these steps separately or in an order preferred by the engineer.

[Example 3.1]
A young design engineer with a machine tool company has the assignment of developing a new concept for the push-pull tool device for releasing the tool from the spindle in machining centers. The device must be functional and simple, and fit spatial constraints of existing machining centers. He has six weeks.

A creative solution to this task might evolve in the following manner.

Preparation phase

The engineer has been in the company for one year, and this is his first major design assignment. He has a strong desire to accomplish this task successfully. He first studies the existing designs in the company. He realizes that before the automatic tool changer extracts the tool from the spindle of a machining center, the tool bar must be pushed to release the tool mechanism by the spindle. And for all existing designs, this function is achieved by hydraulic cylinders. Noise and high cost are the major disadvantages. The engineer then collects available commercial catalogues and service manuals of machining centers in the market, carries out an exhaustive patent search, and requests related reports from allied research institutes and universities. After an intensive and careful study of these materials, he makes an appointment with a senior design engineer in the company, asks him many prepared questions and obtains suggestions regarding this design project. Finally, the engineer decides to come up with such a design: a new push-pull tool mechanism that is simple and reliable in structure, and inexpensive in cost.

Incubation phase

The engineer evaluates available documents and information. He applies almost all the methods and techniques he learned in school, and tries very hard to generate novel solutions. However, he seems to discover that all good solutions have been used and even patented. The young engineer now faces tremendous pressure, and he is very frustrated.

Illumination phase

Four weeks has gone by, the engineer is physically and mentally tired. He then decides to do nothing but sleep over the weekend, subconsciously giving up the assignment. However, something magical happens. He dreams up a rough concept. A device with a cylindrical cam and an elliptic trammel may be a feasible solution.

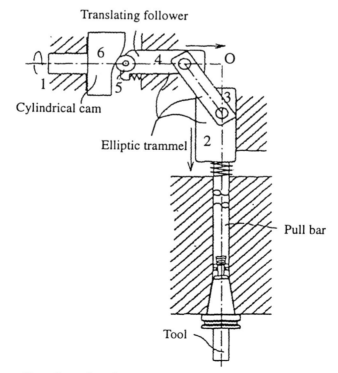

Figure 3.2 **A push-pull tool mechanism**

Execution phase

The engineer then jumps up from bed and sketches this incredible insight immediately. In the final two weeks, he makes the structural skeleton of his idea, develops a computer program to simulate the kinematic, dynamic and stress analysis of the proposed concept, and he proves that this device is feasible. Finally, he presents this dreamed up push-pull tool mechanism, shown in Figure 3.2, to his supervisor. The cylindrical cam (link 6) is driven by a chain (not shown) from the power source to force the translating follower (links 2 and 5) to move along the horizontal direction. The translating follower (link 4), connecting link 2, and slider 2 constitute the elliptic trammel. The tool can then be released by pushing the pull bar that is driven by the elliptic trammel through the cylindrical cam mechanism. This design is simple and reliable in structure. The design requires no hydraulic cylinders as the power input, and is therefore inexpensive.

3.3 Creativity's Characteristics

Everyone is born to have tremendous creative potential, and everyone has some capability for creative thinking. However, creative behavior is a function of the individual personality. Some people have positive creative attributes, and some people have negative attributes. For this reason, it is important to understand the characteristics of creative behaviors, to identify the obstacles of creativity, and to develop the attributes for enhancing imaginative and creative thought.

3.3.1 Creative Person

Children are born with an unlimited potential for creativity. They have natural creative characteristics. They accept disorder, think of alternatives, take risks, and like to play with ideas. They sometimes lag in verbal ability. They are often attracted to the mysterious. Furthermore, children are persistent, playful, emotionally sensitive, and usually have a high energy level.

As children grow up into adults, most of them lose their creative behaviors. Nevertheless, a creative adult can be identified based on his personal characteristics. He is likely to have a driving curiosity to solve problems, an intense faith in himself, a high tolerance for the ambiguous, and a good sense of humor. He habitually accepts chaos and change, does not value job security, can easily accept failure, and prefers to work alone. And a creative adult is usually insensitive to the feelings of others, aggressive in goals, unconventional, intelligent, persistent, energetic, and gullible.

On the other hand, an adult is likely to be uncreative if he possesses the following negativity attributes: resistance to change, desire for conformity, desire for security, desire for organized routine and order, no desire to experiment, competitive jealousy, fear of ridicule, fear of failure, distrust of wild ideas, concern for effect rather than cause, and cynicism.

3.3.2 Barriers to Creativity

Just as certain conditions stimulate creative activity, there are other conditions that depress creative thinking and creative behavior. Thus, though the engineer may have high creative potential and the intellectual ability to synthesize, analyze, and evaluate the problem and the desired solution, he still may not be creative. This may be mainly due to engineer's emotional, cultural, perceptual, and other barriers.

Emotional barriers

An engineer is likely to lose his creative behavior under high emotional tension. Emotional constraints are perhaps more damaging than other types of creative barriers, because they can have a lasting influence upon one's personality. Fear of anything, discouragement by others, distrust of supervisors, suspicion of colleagues, ... etc., may provide an emotional obstacle to creativity. And fear, such as fear of failure, criticism, embarrassment, ridicule, loss of employment, or social disapproval, can reduce the flow of imaginative ideas. Emotional barriers are the most difficult to cope with.

Cultural barriers

Cultural restraints may be insensible, but they are very real. People's habits, emotions, and thoughts are strongly affected by the cultural influences that surround him. At an early age, he learns that his associates disapprove of some of his actions and reward others with accolades and commendation. Such rewards may motivate him to make supreme efforts to gain recognition, but condemnation may make him afraid of deviating from group opinion and thus stifle his creative and imaginative thought. Furthermore, most people tend to conform to the existing pattern of lives. They are reluctant to accept change, and they are either indifferent or negative to propose new ideas. This is why few creative people like da Vinci, Galileo, and Mozart lived to see people accept the products of their imaginations.

Perceptual barriers

Perceptual obstacles also may block the innovative solution of problems. Often it is difficult to visualize remote relationships or to distinguish between cause and effect. Often one does not use all of his senses for observation, but rather choose only the data he wants to use. When one uses only partial information and fails to investigate thoroughly or to define terms, his view of the problem and its solution may be limited.

[Example 3.2]
Which one of the two lines shown in Figure 3.3(a) is shorter?

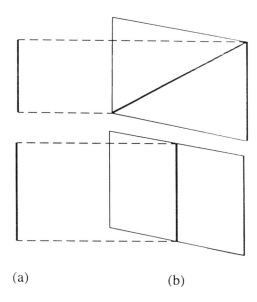

(a)   (b)

Figure 3.3 **An example of perceptual barriers**

A direct answer to this question is the line on the top is shorter. However, if these two lines are not of true lengths the answer might be different. For example, if the lines exist in two dimensional space, as shown in Figure 3.3(b), the answer depends upon the perspective from which the lines are pictured. It is possible that the line at the bottom is shorter.

Other barriers

What is listed below are other types of barriers that limit creative thinking and behavior: false constraints of the problem, failure to recognize all the conditions relating to the problem, failure to get all the facts, failure to investigate both the obvious and the trivial, limited scope of basic knowledge, use of rigid problem-solving strategies, preconception and reliance upon the history of other events, and inclusion of extraneous environmental factors.

[Example 3.3]
Is it possible to draw four straight lines through all the nine dots shown in Figure 3.4(a) without lifting the pencil and retracting a line?

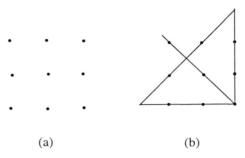

(a)   (b)

Figure 3.4   **An example of false constraints**

When looking at the problem, most people assume a false constraint that the boundaries are confined to the dots. Yet such a restriction is unstated and unnecessary. When this assumed boundary condition is ignored, a solution is not difficult to find as shown in Figure 3.4(b).

The performance of most existing designs have been improved step by step throughout the long history of the products, especially mechanical products. They are on the way to reaching the technology ceiling, and available space for further improvements is quite limited. For example, an engineer is assigned to design eyeglasses that are lighter than existing products. If his thinking domain is bounded by trying to change the material of the lenses and the frame of eyeglasses, the improvement should be small. However, the inventor of contact lenses, who walked out of the tradition of the design, created a new domain for solving the problem.

A highly creative engineer should seek new methods and/or directions for solving old problems of traditional products. In the mechanical history of reducing friction between two members with relative motion, lubricants are used for direct surface contact, Figure 3.5(a); then ball bearings are designed to provide rolling contact, Figure 3.5(b); then air bearings are invented, Figure 3.5 (c); then electric-magnetic means are applied, Figure 3.5(d). Each innovative concept provides a quantum improvement in the problem solution.

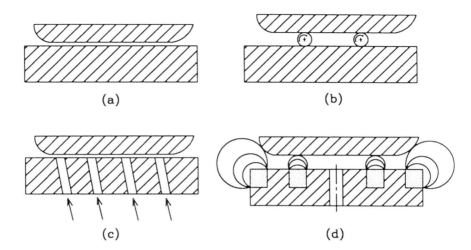

Figure 3.5 Concepts for reducing friction

3.3.3 Creative Enhancement

Most creative ideas occur through a slow and deliberate process that can be enhanced with experience, study, and practice. Furthermore, there are some specific actions and attitudes that can be employed to overcome obstacles to creative thinking. A considerable literature has accumulated on creative enhancement, but the steps given here encompass most of what has been suggested.

Be motivated

An engineer must be motivated to use imaginative and innovative thought. Motivation is the power source that drives all engineers forward in their role as problem solvers, innovators, and creators.

Be confident

Engineers must develop confidence that they can provide a creative solution to a problem. Although one may not visualize the complete picture to the final solution at the beginning of the design process, one must have self-confidence; one must believe that a solution will develop before the work is finished. Since the feeling of successes provides confidence, it should be a good approach to start and build one's confidence up with small successes.

Be patient

Much patience is necessary to lead to creativity. The creative process usually requires relatively long periods of time with high frequency of failure. Inadequate solutions are to be expected. A continuous willingness to try again is important. Furthermore, creativity often requires hard work. Most problems will not succumb to the first attack. They must be pursued with persistence. After periods of intensive concentration, one should allow time for the necessary incubation. To quote Edison's famous comment: "Invention is 95 percent perspiration and 5 percent inspiration."

Be open minded

Having an open mind means being receptive to ideas from all possible sources and accept for a variety of problem solving strategies. The solutions to problems are not the property of a particular discipline. One should consider approaches that might be used by other disciplines. One should try to approach problem situations with a clear mind stimulated but not restrained by experiences. One should look for relationships that are remote and solutions that are unusual and nontraditional. One should avoid placing unnecessary constraints on the problem being solved. And one should search for different ways to view the problem.

Open imagination

Imagination is more important than knowledge, for knowledge is limited, whereas imagination embraces the entire world. An engineer must give rebirth to the vivid imagination he had as a child. One way to do so is to ask "why" and "what if" again, even at the risk of displaying a bit of naivete.

Suspend judgment

Although creative ideas develop slowly, nothing inhibits the creative process more than critical judgment of an emerging idea. Engineers, by nature, tend toward critical attitudes, so special forbearance is required to avoid judgment at the early stage of problem solving.

Set problem boundaries

Engineers usually greatly emphasize on proper problem definition in the process of problem solving. Establishing proper boundaries of the problem is an essential part of problem definition. This, in fact, does not limit creativity, but rather focuses it. Otherwise, the direction of the solution might not converge and the number of possible solutions might go wild.

Decompose the problem

If the target problem is too complex a system, the engineer should divide the problem into manageable subsystems and concentrate on solving one subsystem at a time. Once solutions to subproblems are generated, the engineer then assembles these solutions into a complete one.

3.4 Summary

Innovation is the introduction of a new idea, method, or device. Creativity is a defining of previously unknown things, it is innovation to meet a need. Invention is a process to produce something useful, and it is the result of creative thought.

The creative process may include preparation phase, incubation phase, illumination phase, and execution phase. However, not all of the steps in each phase of the creative process occur in the solution to every problem.

The human mind has an unlimited potential for creative thinking. Our ability to perform creative thought is dulled by a lack of mental exercise and is often suppressed by emotional, cultural, perceptual, and other types of barriers. These barriers can be overcome through the use of special techniques.

In order to enhance creativity in problem solving, an engineer should be confident, patient, open minded, and imaginative. He should suspend judgment, set problem boundaries, and decompose a complicate problem.

An engineer should recognize his creative potential and should lift the barriers to creative thought before he uses any creative techniques or methodologies to unleash his creativity for design projects.

Problems

3.1 Give an example to explain the difference between innovation and invention.

3.2 Provide a creative design example of your own, by someone else, or from literature, and describe the creative process of this novel idea.

3.3 Identify the creative characteristics of a kindergarten child in your family or neighborhood.

3.4 Discuss with four of your classmates to identify the characteristics of the person whom you think is most creative in the class.

3.5 What are your major emotional barriers to creativity?

3.6 Provide an example of perceptual barriers to creative thought.

3.7 Identify an import product that has cultural barriers to creative thought.

3.8 Identify your negative creative attributes.

3.9 What attitudes should you develop to enhance your creative thinking abilities?

References

Beakley, G. C. and Leach, H. W., Engineering - An Introduction to A Creative Profession, Macmillan, 1967.
Cross, N., Engineering Design Methods, John Wiley & Sons, 1994.
Dieter, G. E., Engineering Design, McGraw-Hill, 1983.
Edel, Jr., D. H., Introduction to Creative Design, Prentice Hall, 1967.
French, M., Invention and Evolution, Cambridge University Press, 1994.
Hill, P. H., The Science of Engineering Design, Holt, Rinehart and Winston, 1970.
Krick, E. V., An Introduction to Engineering and Engineering Design, John Wiley & Sons, 1965.
Kuo, T. C., A Study on the Automatic Tool Change Mechanisms of Roller-Gear Cam Type, Master thesis, Department of Mechanical Engineering, National Cheng Kung University, Tainan, Taiwan, June 1997.
Lumsdaine, E. and Lumsdaine, M., Creative Problem Solving, McGraw-Hill, 1995.
Pearson, D. S., Creativeness for Engineers, Edwards Brothers, 1961.
Rubinstein, M. F. and Pfeiffer, K. R., Concepts in Problem Solving, Prentice-Hall, 1980.
Shoup, T. E., Fletcher, L. S., and Mochel, E. V., Introduction to Engineering Design, Prentice-Hall, 1981.
Vidosic, J. P., Elements of Design Engineering, Ronald Press, 1969.
Von Fange, E. K., Professional Creativity, Prentice-Hall, 1959.
Webster's New Collegiate Dictionary, G. & C. Merriam, 1981.
Wright, P. H., Koblasz, A., and Sayle, W. E., Introduction to Engineering, John Wiley & Sons, 1989.

CHAPTER 4

RATIONAL PROBLEM SOLVING

Traditionally, there are rational methods that encourage a systematic approach to design. These methods are intended to widen the search space for potential solutions and/or to smoothen group decision-making. This chapter describes several important rational methods for problem solving, such as analyzing existing designs, information search, and checklist.

4.1 Analysis of Existing Designs

Generally speaking, a solid review of available existing design solutions stimulates ideas for new designs. Knowledge of existing solutions may be obtained either analytically or experimentally by studying competitive products. In the early stage of designing a product, especially mechanical devices, for finding a feasible initial concept, a systematic exploitation of proven ideas is particularly helpful.

For example, if a design engineer plans to develop a new rapier type shuttleless loom, he can start by analyzing the function and performance of an available commercial product such as the one shown in Figure 4.1. The transmission mechanism of this design consists of an oscillating variable pitch lead screw as the output member, a reciprocating slider along the axis of the screw as the input member, four conical meshing elements adjacent to the screw and the slider, and the frame of the machine. Figure 4.2 shows its kinematic diagram.

50 ▪ 4. Rational Problem Solving

Figure 4.1 An existing rapier type shuttleless loom (Courtesy of I-Chin Company) (see Appendix)

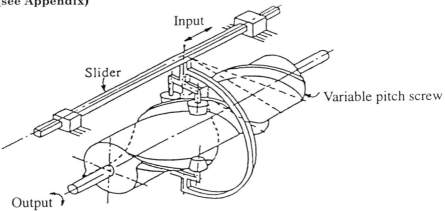

Figure 4.2 Transmission mechanism of the rapier type shuttleless loom

4.1.1 Mathematical Analysis

Mathematical analysis is a powerful tool for learning how to improve existing designs. A thorough analysis of an available design will reveal fundamental knowledge about its function and performance. This knowledge exposes shortcomings that lead to ideas for improvements. These ideas, in turn, require more analysis that provides more fundamental understanding of a design, leading to further ideas for possible improvements. Due to the increasing performance of computers, and with the help of numerical techniques and optimal methods, analytical solutions of design problems have become a powerful means of arriving at an optimum design.

Take, for example, the variable pitch screw of the rapier type shuttleless loom shown in Figure 4.2. A stress analysis is carried out to realize the strength and rigidity of the screw due to various loading. Figure 4.3 shows the geometric and finite element mesh models and the contact stress analysis of the screw by using PATRAN and MARC in a workstation. The result proves that this design is strength and rigidity safe. Furthermore, mathematical expressions of geometric surfaces of the screw are derived based on the coordinate transformation method, differential geometry, and enveloping theory. Figure 4.4 shows the geometric solid modeling of the variable pitch screw for kinematic simulation.

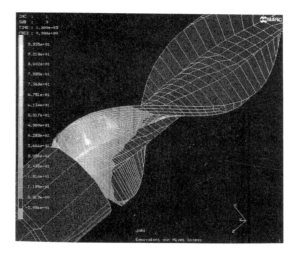

Figure 4.3 Contact stress of the variable pitch screw (see Appendix)

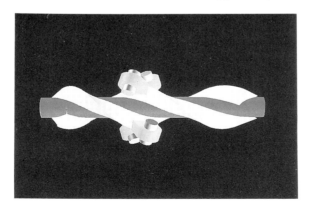

Figure 4.4 Geometric solid modeling of the variable pitch screw (see Appendix)

4.1.2 Experimental Tests and Measurements

Experimental tests are needed mainly for the following purposes:
1. To validate an ideal mathematical model.
2. To understand the problem and find out what is happening physically.
3. As empirical means, if the problem is too complicated to analyze mathematically.

And measurements are necessary for obtaining important data of a design.

Experimental tests and measurements on existing products are among the designer's most important sources of information. Design can be interpreted as the reversal of physical experiment. In mechanical industries, experimental investigations are an established means of arriving at solutions. This approach has organizational repercussions, since in the creation of such products, experimental development is often incorporated within the design activity.

For the variable pitch screw of the rapier type shuttleless loom shown in Figure 4.2, an analytical algorithm is derived to compare the deviation between the point coordinates of the machined screw surface and theoretical screw surface. An off-line measurement technique is developed on a 4-axis coordinate measuring machine (Figure 4.5). Furthermore, a testing rig is set up for measuring the angular velocity of the crank, the displacement, velocity and acceleration of the screw, the input power, and the shaking forces (Figure 4.6). And it is found that the problem of shaking forces can be greatly improved (Figure 4.7) by adding balancing counterweights.

4.2 Information Search

In addition to a review of existing ideas and designs, several information sources, such as literature, patents and experts, are worth consulting for idea stimulation.

4.2.1 Literature Search

Fundamental theories and technological data of a target problem provide a wealth of important information for design engineers. Such information can be found in text and reference books, academic journals, research dissertations, technical reports, trade magazines, commercial product brochures, manufacturers' application literature, publications of professional societies, patent files, and in-house research notebooks. They provide a most useful and basic source of known solution possibilities, and sometimes even stimulate new ideas.

Figure 4.5 Precision measurements of the variable pitch screw (see Appendix)

Figure 4.6 Dynamic testing of the variable pitch screw transmission mechanism (see Appendix)

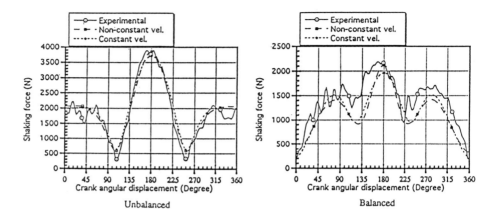

Figure 4.7 **Dynamic improvements of the variable pitch screw transmission mechanism**

For the variable pitch screw transmission mechanism of the rapier type shuttleless loom shown in Figure 4.2, an exhaustive literature search uncovered more than one hundred research papers.

4.2.2 Patent Search

A patent search for the target problem is always necessary and valuable. The patent system is a rich source of ideas for practically any design problem. An understanding of existing patents is needed to avoid the misuse of legal intellectual properties. Sometimes, patents may provide some important insights regarding the topic of interest. Many good references on the patent system and its applications are available.

It is entirely practical and legal to use a patented design concept if one changes the idea sufficiently to avoid infringement. With ingenuity, the design can often be modified for use without patent violation. Furthermore, expired patents are very valuable sources of ideas for alternative designs. One should not feel defensive about basing one's design on an expired patent. The very reason an inventor is given exclusive right to his invention is that it will be available for the public to copy after the patent has expired.

Figure 4.8 shows an existing patent relating to the variable pitch screw transmission mechanism of the rapier type shuttleless loom shown in Figure 4.2. A careful study of the literature description of this patent provides a quick and valuable lesson regarding the design know-how of this machine.

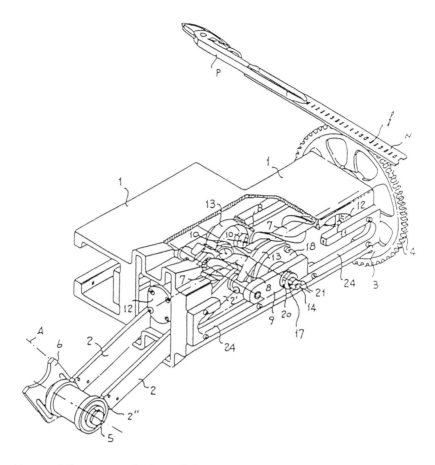

Figure 4.8 **A pattern of the transmission of the rapier type shuttleless loom (U.S patent No.4,052,906)**

4.2.3 File of Experts

A design engineer is normally a generalist, one who puts together a group of components into a rational whole to accomplish a given purpose. As a result, he usually works across a number of engineering disciplines and needs to call on experts in special fields to advise on specific design approaches. Experts, including local experts in the company, experts in consulting companies, experts in research institutes and universities and others, are usually most helpful in identifying problem areas within a design concept and in supporting careful evaluation.

4.3 Checklist Method

The checklist is the simplest kind of rational method for problem solving. It is an accumulation of points, areas, or possibilities that serve to provide hints by checking off the items on a prepared list against the problem. In design terms, a *checklist* is a set of questions that can be asked about any project in an attempt to stimulate ideas. It may be a list of questions to be asked in the initial stages of design, a list of features to be incorporated in the design, or a list of criteria that the final design must meet. The checklist is used by many engineers as a group effort to identify different ways of viewing a problem.

During the early design phases of an engineering project, general listings of questions regarding the design subject are useful to avoid the omission of important attributes and to suggest possible improvements. The checklist method externalizes what one has to do, so that one does not have to try to keep it all in one's mind and so that one does not overlook something. It formalizes the process by making a record of items that can be checked off as they are collected or achieved, until everything is complete.

4.3.1 Checklist Questions

The development of checklist questions concerning the general area of the problem is perhaps the most common checklisting approach. Its purpose is to point out other ways of looking at a problem and thus to stimulate creative thinking. By asking oneself through providing questions, one opens up the opportunity to investigate the side effects of the problem. The more questions asked, the more unusual features explored, the more likely that new and innovative ideas may occur.

A typical checklist includes the following kinds of information:
1. Physical attributes, such as shape, size, weight, position, velocity, acceleration, force, torque, power, efficiency, friction, vibration, noise, pressure, temperature, ... etc.
2. Variations in functional aspects, materials, packaging, applications, manufacturing processes, ... etc.
3. Characteristics of shape, finish, details, appearance, feel, fashions, maintenance features, assembly methods, energy sources, ... etc.
4. Social aspects, such as timing, cost, recycling, availability, human compatibility, degree of complexity, ... etc.
5. Possible rearrangements, recombinations, modifications, and elimination of excessive details and features.

Each question on the list may bring to mind other questions that might be raised. Once the checklist is made, it is helpful to ascertain which of the questions are of primary importance and which are of secondary importance to the solution of the problem.

[Example 4.1]
A motorcycle company, which used to produce 50 cc and 90 cc scooters, has decided to develop 250 cc motorcycles. A senior engineer is responsible to design a new transmission for the project.

The senior design engineer may start the project, based on the checklist method, by providing the following list of questions for his team:
1. Could the currently used transmissions in 50 cc and 90 cc scooters be modified for this application?
2. Should the new transmission for 250 cc motorcycles be larger and stronger than the present one?
3. Could the present transmissions in scooters be magnified for this purpose?
4. Could new technology be combined with the present design to meet the needs?
5. Is there new technology that could be used to design a totally new transmission?
6. Could the newly designed transmissions for 250 cc motorcycles be used for other purposes?

4.3.2 Checklist Transformations

Another type of checklist is used to stimulate ideas based on the process of transformations such as adaptation, combination, magnification, minification, modification, rearrangement, reverse, substitution, and put to other uses.

Adapt?
The solution of a problem in one field is applied to a similar problem in another field.

Typical checklist questions are: What can be adapted for use? What can be copied for use? What can be modified for use? What else is like this? What other idea does this suggest?

For example, the design of the tail wing of sporty cars is adapted from the concept of the aerodynamic force on the wings of aircraft (Figure 4.9). And the high tech appearance of the front suspension of a motorcycle shown in Figure 4.10 is adapted from the nose landing gear of aircraft.

Combine?
The solution of a problem is obtained by combining the solutions of different problems.

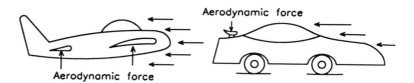

Figure 4.9 Adaptation of aerodynamic force on passenger vehicles

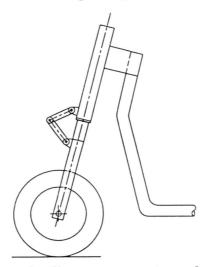

Figure 4.10 Adaptation of aircraft nose landing gears on motorcycles

Typical checklist questions are: Combine purposes? Combine ideas? Combine principles? Combine methods? Combine groups? Combine units? Combine components? Combine hardware? Combine appeals? Combine materials? How about a blend, an assortment, an ensemble?

For example, a composite material is the combination of different materials, and a modern fax machine has the combined functions of fax, telephone, copier, and even answering machine. Another example of combining purposes is presented as follows. The velocity ratio of the output member to the input member for a gear train is a constant and that for a four-bar linkage is not. However, by combining the gear train into the four-bar linkage as shown in Figure 4.11(a), the design requirement that the velocity ratio in a certain period is a constant, Figure 4.11(b), can be achieved.

<u>Magnify?</u>

The solution of a problem is obtained by enlarging the fact, the appearance, or the significance of an existing design.

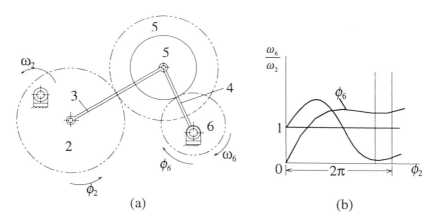

Figure 4.11 **Combination of a gear train and a four-bar linkage**

Typical checklist questions are: What to add? Larger? Higher? Wider? Longer? Thicker? Heavier? Stronger? Greater? Duplicate? Multiply? Exaggerate? Extra value?

For example, the anti-dive front suspension mechanism shown in Figure 4.12 for an 125 cc scooter is obtained by enlarging the design of a 50 cc product.

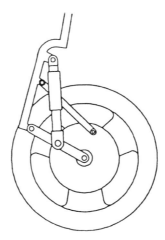

Figure 4.12 **Magnification of a scooter anti-dive suspension mechanism**

Minify?

The solution of a problem is obtained by lessening the fact, the appearance, or the significance of an existing design.

Typical checklist questions are: What to subtract? Smaller? Lower? Shorter? Narrower? Lighter? Weaker? Condensed? Miniature? Omit?

For examples, the design of the windshield wipers for subcompact cars is normally a smaller version of the wiper systems of existing mid-size cars, and notebook computers are the minification of desktop personal computers.

Modify?

The solution of a problem is obtained by making minor changes to an existing design.

Typical checklist questions are: Change motion? Change contours? Change shape? Change form? Change weight? Change color? Change sound? Change odor? Change description? Change meaning? Other changes?

For example, the design of the transmission mechanism with canonical meshing elements shown in Figure 4.2 is a modification of the patented design with cylindrical meshing elements shown in Figure 4.8.

Rearrange?

The solution of a problem is obtained by the adjustment of the components or subsystems of an existing design into a proper order, a suitable sequence, or relationship.

Typical questions are: Change schedule? Change sequence? Change order? Change layout? Change pace? Change place? Change pattern? Repackage? Piece together differently? Interchange components? Transpose cause and effect?

Figure 4.13(a) shows the slider-crank mechanism of a press where link 1 is the frame, link 2 is the input crank, link 3 is the connecting rod, and link 4 is the output slider. If this mechanism is rearranged as the one shown in Figure 4.13(b) with link 1 as the output slider, link 2 as the input rocker, and link 4 as the frame, it is a hand pump mechanism.

Reverse?

The solution of a problem is obtained by operating, arranging, or acting an existing design in a manner contrary to the usual.

Typical checklist questions are: Up instead of down? Positive instead of negative? Inside instead of outside? Backward instead of forward? Reverse roles? Opposite pattern? Opposite sequence?

Dining room lights that throw a beam of light at the ceiling is an example of reverse. And, for the slider-crank mechanism shown in Figure 4.13(a), if link 4 is the input and link 2 is the output, it is a slider-crank mechanism for an engine.

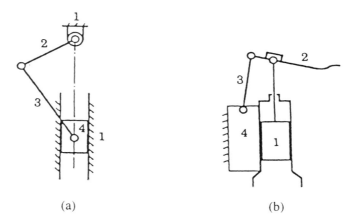

Figure 4.13 Rearrangement of slider-crank mechanisms

Substitute?

The solution of a problem is obtained by replacing the components or subsystems of an existing design with something else.

Typical checklist questions are: What other power, parts, materials, process, principle, theory, or method can be used? Who else instead? What else instead? Light instead of dark? Round instead of square? Other material? Other ingredient? Other place? Other time? Other process? Other power source? Other approach?

For example, an electric motor may be used to substitute the two-cycle engine in a lightweight scooter, as shown in Figure 4.14, for avoiding the problem of air pollution.

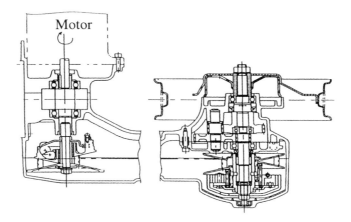

Figure 4.14 Substitution of an electric motor for the two-stroke engine in scooters

Put to other uses?

The solution is considered for other uses.

Typical checklist questions are: What other uses does it have in its present form? What other uses are there if the idea is modified? Can it perform a function that was not originally intended? New ways to use as is? What could be made from this?

For example, the concept of the bed warmer invented in the Han Dynasty of China in the first century is the same as the modern gyroscope, Figure 4.15.

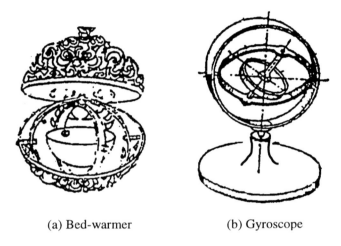

(a) Bed-warmer　　　　(b) Gyroscope

Figure 4.15　An ancient bed warmer and a modern gyroscope

4.4 Summary

In the process of generating new concepts for a design task, an engineer normally starts by applying traditionally rational methods for problem solving. He may analyze available existing designs mathematically and/or experimentally to obtain ideas for new designs. Engineering analysis using mathematical models often leads to discovery of improved design concepts. Design characteristics of existing devices or products can be understood through physical experimental tests and measurements. Furthermore, an engineer may search for literature, patterns, and files of experts regarding the subject to accumulate ingredients for idea initiation. He may apply the method of checklisting by providing a list of questions to guide his thinking for solution generation. However, solving problems based on rational methods carries the danger of causing the designer to stick with known solutions instead of pursuing new paths.

After the design engineer has searched all the sources available for alternative designs of the target problem based on rational methods, he may decide that a basically new approach is desirable. He may wish to generate a design that is superior to those he has found. He may also wish to come up with a new design that permits a patent protection. Any of these is a proper motivation for the design engineer to try to invent a new device, product, system, or process to meet the specifications. To prepare for such type of activity, creative techniques for stimulating invention will be discussed in Chapter 5.

Problems

4.1 Explain how mathematical tools can play a role in analyzing the performance of a mountain bike.
4.2 Discuss how to obtain performance data of a motorcycle based on the techniques of experimental tests and measurements.
4.3 Identify at least three books regarding engineering creativity.
4.4 Search at least five papers regarding variable pitch screws.
4.5 Search at least three patents regarding bicycle derailleurs.
4.6 Provide a list of questions for improving students' campus parking.
4.7 Repeat Problem 4.6 with a group of three.
4.8 Answer the question "What if gravity were to disappear for ten minutes each day at noon?" with a group of five.
4.9 Give an example for each of the checklist transformations.

References

Alger, J. R. M. and Hays, C. V., Creative Synthesis in Design, Prentice-Hall, 1964.
Cross, N., Engineering Design Methods, John Wiley & Sons, 1994.
Edel, Jr., D. H., Introduction to Creative Design, Prentice Hall, 1967.
Hill, P. H., The Science of Engineering Design, Holt, Rinehart and Winston, 1970.
Middendorf, W. H., Engineering Design, Allyn and Bacon, 1969.
Pahl, G. and Beitz, W., Engineering Design, Springer-Verlag, 1977.
Shoup, T. E., Fletcher, L. S., and Mochel, E. V., Introduction to Engineering Design, Prentice-Hall, 1981.
Wright, P. H., Koblasz, A., and Sayle, W. E., Introduction to Engineering, John Wiley & Sons, 1989.
Yan, H. S., "Design and manufacturing of variable pitch screw transmission mechanisms," Proceedings of International Conference on Mechanical Transmissions and Mechanisms, Tianjin, China, July 1-4, 1997, pp. 17-20.

CHAPTER 5

CREATIVE TECHNIQUES

There are some creative techniques that can be applied to achieve a number of potential design solutions, either individually of by a group effort. Each technique provides a logical step by step procedure to initiate and/or generate solutions for the target problems. This chapter introduces creative techniques that have proven most useful in stimulating engineers' production of ideas. These are attribute listing, morphological chart analysis, and brainstorming.

5.1 Introduction

History has shown that the majority of creative and innovative effort in industry is the result of individual thought processes. In most cases, an engineer has the responsibility for a design problem that is to be solved, and the solution may involve the activities of many people with various backgrounds. However, original ideas can usually be traced to an individual rather than to a group. The major group function is to amplify or modify the ideas first conceived by an individual.

For example, Figure 5.1 shows a novel design concept for a hybrid transmission, by combining an existing continuous variable transmission and an available planetary gear train, for two-stroke motorcycles to improve the overall performance in terms of mechanical efficiency. This idea was initiated by a professor and a graduate student in an university. And the final design, as shown in Figure 5.2, was completed through gradual improvements by discussions of the research group in the university and engineers in industry.

66 ▪ 5. Creative Techniques

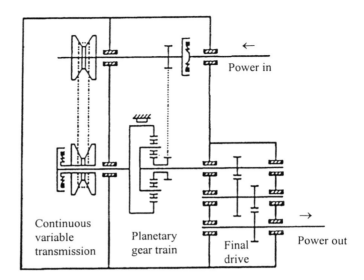

Figure 5.1 **Initial concept of a hybrid motorcycle transmission**

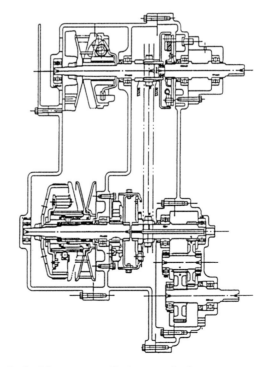

Figure 5.2 **Final design of a hybrid motorcycle transmission**

Group creative effort on design problems is probably most effective during the early stages of important or complicated projects, in order to stimulate the development of ideas for design concepts. One idea may lead to another idea, which in turn forms a foundation for yet other ideas. Such group activity assures that the final design concepts obtained are broad based.

In the following sections, individual creative techniques such as attribute listing and morphological chart analysis are described first. Then, brainstorming, a group creative technique, is explained.

5.2 Attribute Listing

An *attribute* is an inherent characteristic. It is an object closely associated with or belonging to a specific thing. Professor Robert Crawford of the University of Nebraska, in 1954, used the term *attribute listing* for the systematic searching for variations on each of the main attributes of a design. The objective of attribute listing is to focus one's mind on the basic problem and to stimulate one's thinking process for generating new concepts that might better solve the problem at hand.

Attribute listing is an individual creative activity. Different people may interpret the same problem differently and may produce different lists of attributes.

5.2.1 Procedure for Attribute Listing

The working steps of attribute listing technique can be summarized as follows:
1. List the major attributes of an idea, device, product, system, or principal parts of the problem.
2. Change or modify all listed attributes, no matter how impractical, for possible improvements that can be made for the target idea, device, product, system, or principal parts of the problem.

It is helpful to combine this creative technique with the rational checklist method for concept generation.

5.2.2 Examples

Two examples are given to show how the attribute listing technique could be used to improve the design of fax machines and mechanical push button locks.

[Example 5.1]
Fax machines.
1. The major attributes of fax machines are:
   (a) Function of machine.
   (b) Type of paper.
   (c) Size of paper.
   (d) Shape of machine.
2. Ideas of each attribute are:
   (a) Additional function of machine: telephone, copy, recording, radio, alarm.
   (b) Type of paper: plain paper, special paper, transparency.
   (c) Size of paper: A4, A3, B4, B3, pocket size, adjustable.
   (d) Shape of machine: oval, rectangular, round, triangular.

[Example 5.2]
Mechanical push button locks.

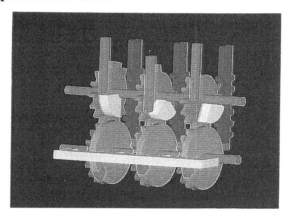

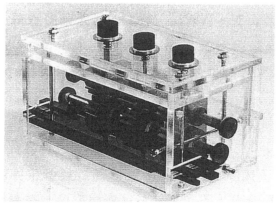

Figure 5.3 **A new sequential push button lock with variable passwords** (see Appendix)

1. The major attributes of most existing mechanical push button locks are:
   (a) Type: one step push button.
   (b) Password: fix number.
   (c) Shape: rectangular.
2. Ideas of each attribute are:
   (a) Improve the product by achieving the function with sequential push button, repeatable push button, and/or multistep push button.
   (b) Improve the product by providing the function with variable passwords.
   (c) Improve the product by producing various shapes and even colors of the locks with industrial design.

Figure 5.3 shows a new concept of sequential push button lock with variable passwords by computer simulation and by an acrylic model.

5.3 Morphological Chart Analysis

Morphology means the study of structure or form. A *morphological analysis* is a systematic approach to analyze the structure or form of an idea, object, device, product, system, or process. A *morphological chart analysis* is a summary of this analysis. It takes the technique of attribute listing one step further towards abstraction and greater variety.

Morphological chart analysis is a systematic way of examining all possible combinations of known variables of available problem solutions. It involves the development of a list of major independent parameters associated with the problem, along with several design alternatives for each parameter. The intent is to clear up a fuzzy problem situation and to uncover combinations of components that would not ordinarily develop from the normal process.

5.3.1 Characteristics of Morphological Chart Analysis

The advantage of morphological chart analysis is the short time needed to complete a matrix for problem solving. The main difficulty is that of identifying a set of functions that are: essential to any solution, independent of each other, inclusive of all parts of the problem, and few enough in number for producing a matrix that can be searched in a short time.

Morphological chart analysis works best when the problem can be readily decomposed into subproblems. Each subproblem should represent a meaningful part of the major problem. Its success in stimulating ideas depends largely upon the engineer's identifying all of the significant parameters that affect the design.

A morphological chart displays the complete range of possible subsolutions that can be combined to make a solution. The number of possible combinations is usually very high. The resulting solutions include not only existing designs but also a wide range of variations and completely novel solutions.

The creative technique of morphological chart analysis can best be applied by individual effort. It enables the investigation of many different combinations of variables that the checklist method does not permit.

5.3.2 Procedure for Morphological Chart Analysis

The necessary steps for morphological chart analysis are:
1. Define major design parameters of the idea, object, device, product, system, or process.

   It is important that the chosen design parameters (features or functions) should be at the same level of generality, and they should be reasonably independent of each other. These design parameters in combination should meet product performance requirements. However, the list of parameters must not be too long, so as to avoid unmanageably large possible combinations of subsolutions. About four to seven design parameters would make a manageable list.
2. List several subsolutions for performing each design parameter.

   These subsolutions can include not only the existing subsolutions of the particular design, but also new ones that might be feasible. The overall problem solution is obtained by combining one subsolution from each design parameter.
3. Set up the morphological matrix.

   Set up a matrix having each major design parameter as one axis of a rectangular array. And each subsolution for performing a major design parameter is considered a coordinate along the dimension of the major design parameter.
4. Identify feasible solutions.

   By selecting one subsolution at a time from each row of the matrix, all the theoretically possible different solution forms for the design are obtained. If the total number of possible solutions is not too large, each potential solution should be considered.

Figure 5.4 shows a typical morphological chart. This chart permits a comparison of different characteristics that might not otherwise be considered. For example, combination $A_3B_2C_4D_2$ might be a feasible solution or it might prove to be an impossibility.

5.3. Morphological Chart Analysis

Design Parameters	Alternative Subsolution			
A	A_1	A_2	A_3	A_4
B	B_1	B_2		B_3
C	C_1	C_2	C_3	C_4	C_5
D	D_1	D_2	D_3	D_4

Figure 5.4 A typical morphological chart

5.3.3 Examples

Examples are given to illustrate how the technique of morphological chart is applied for creative problem solving.

[Example 5.3]
Design a planar linkage type horizontal tail control mechanism for an aircraft.
1. Major independent design parameters are:
   (a) Type and number of independent inputs.
   (b) Type of joints.
   (c) Number of links.
2. Possible subsolutions for each design parameter are:

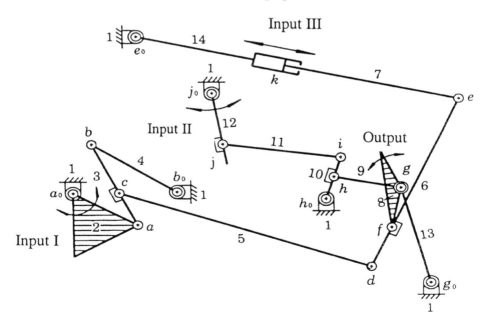

Figure 5.5 An aircraft horizontal tail control mechanism with three inputs

(a) Type and number of independent inputs: 1 input (control stick), 2 inputs (control stick and flap, control stick and stability argumented system), 3 inputs (control stick, flap, and stability argumented system).
(b) Type of joints: revolute pairs only, revolute and prismatic pairs.
(c) Number of links: 10-bar, 11-bar, 12-bar, 13-bar, 14-bar.
3. The combination of these conditions yields various possible solutions available for further evaluation.

Figure 5.5 shows the schematic drawing of the horizontal tail control mechanism of a subsonic attack trainer aircraft. This is a 14-bar planar linkage with 17 revolute joints, 1 prismatic joint, and 3 independent inputs.

[Example 5.4]
Design a power train system for a lightweight motorcycle.
1. Major independent design parameters are:
   (a) Type of power source.
   (b) Type of transmission.
   (c) Type of mechanism of the transmission.
   (d) Type of gear ratio of the transmission.
2. Possible subsolutions for each design parameter are:
   (a) Power source: gasoline engine, turbo engine, electric motor, Stirling engine.
   (b) Transmission: manual, automatic, hybrid.
   (c) Mechanism: gears, belts, chains, linkages, hybrid.
   (d) Gear ratio: three-speed, four-speed, continuous, others.
3. The morphological chart is shown below.

Design Parameters	Alternative Subsolutions			
Power source	Gasoline engines	Turbo engines	Electric motors	Stirling engines
Transmission	Manual	Automatic		Hybrid
Mechanism	Gears	Belts	Chains	Linkages	Hybrid
Gear ratio	Three-speed	Four-speed	Continuous	Others

4. The total combination yields 240 possible solutions. If the power source is limited to gasoline engines and electric motors, the transmission is limited to manual and hybrid, the mechanism is limited to gears, belts and hybrid, and the gear ratio is limited to continuous and others, only 24 solutions are available. If the power source is limited to electric motors, the transmission is limited to automatic, and the mechanism is limited to gears and hybrid, only 8 solutions are available.

5.4 Brainstorming

Brainstorming is perhaps the best known group technique for stimulating and generating new ideas. *Brainstorming*, meaning using the brain to storm, was first introduced by Dr. Alex F. Osborn in 1939. Its purpose is to stimulate a group of people to quickly suggest and produce a large number of alternative ideas from the unrestrained responses of the group.

5.4.1 Characteristics of Brainstorming

In general, any problem that can be stated simply and directly seems appropriate to brainstorming. However, brainstorming is most effective when it is applied to specific rather than general problems. The problem should be limited in scope, open ended, capable of being handled by verbal rather than graphical or analytical means, and familiar to the participants. Problems having only one correct solution or only a few sensible alternatives are not suitable for brainstorming.

Brainstorming can be used at any stage in designing, either at the beginning before the problem has been fully understood or later on when complicated subproblems arise. However, it is usually more useful in the early stages of the design than in the final stages.

A successful brainstorming session may result in one or two exceptionally useful ideas being suggested. However, follow up developments are usually difficult to achieve, since the chain reaction of group procedure results in the combined suggestions of several individuals. No one feels strongly committed to any of the results. Individual motivations are submerged in the group emphasis. When the results are subjected to analyses and criticisms later, there is no one with enough personal involvement to defend and/or develop them further for applications. Brainstorming thus avoids most of the steps in the creative thinking procedure of individuals.

Like other group techniques, brainstorming requires skill and practice to produce successful results, by breaking any prior emotional barriers among the participants.

Group brainstorming is not meant to replace individual effort. It is used solely to supplement individual concept generation. Furthermore, brainstorming can be used either by groups or individually. The same rules are followed except that, in a group, emphasis is given to building on the ideas of others to have a snowball effect. While it has been found that groups produce more ideas than the average individual, it has not been proved that the number of feasible and excellent ideas is larger for groups than for individuals.

Brainstorming cannot be expected to produce ready made solutions for engineering purposes, because problems are generally too complex, difficult, vast, vague, or controversial to be solved by spontaneous ideas alone. Most of the ideas expressed will not be technically or economically feasible, and those that are will often be familiar to the experts. However, if a brainstorming session should produce one or two useful innovative ideas, or even some hints as to in what direction the solution might lie, it will have already achieved a great deal.

5.4.2 Procedure for Brainstorming

All that brainstorming needs is a group of people who are familiar with the general area of the problem. The leader simply states the problem and the other participants say whatever comes to their minds. And, finally, the generated ideas are carefully evaluated.

The procedure for brainstorming can be divided into the following nine steps, in four stages:

Brainstorm group
1. The group leader
2. The group
3. The reporter

Brainstorm session
4. Problem statement
5. Idea brainstorming
6. Idea recording

Brainstorming evaluation
7. Individual review
8. Group review

Brainstorming report
9. Final report

For an important problem, preliminary analyses may be carried out before reconvening the group for the next stage. However, on less complex problems, where evaluation is straightforward and effective judgment may be based on experience and knowledge, the whole process may be completed at one meeting of the group.

In what follows, each stage of brainstorming is described in detail.

5.4.3 Brainstorming Group

Assume that a design engineer has a problem to solve and he seeks group help in conceiving potential design concepts. Normally, he takes charge of the group session planning and execution. He acts as the group leader, selects group members, and appoints one or two reporters to record the brainstorming session.

Group leader

In a brainstorming session, the leader moderates the group to suggest solutions to a stated problem. With the right kind of leadership, the group can accomplish a great deal.

Before beginning the session, the group leader should review the rules of brainstorming with the participants. Then he opens the session by providing each panel member with a brief problem statement. He might continue the session by expressing a few absurd ideas or mentioning an example from another brainstorming session. During the session, the group leader must keep the atmosphere free and easy, ensure that no one criticizes the ideas of other participants, stop any member who offers an evaluation, and encourage the flow of new ideas whenever the productivity of the group slackens. Furthermore, the group leader should never lead in the expression of ideas.

The group

A brainstorming group should be formed according to the following guidelines:
1. The group is usually composed of from five to ten persons.

   Fewer than five provide too small a spectrum of opinion and experience, and hence produce too few stimuli. With more than ten, close collaboration may decline because of individual inertia and cultural barriers.
2. The group should not be confined to experts.

   The group should not just be experts or those knowledgeable in the problem area. Ideally, the participants should have different professional backgrounds and not be involved in any strong professional rivalries or competition. With such a diverse group, their roles in the solution process may be different; but they share the common objective of reaching an optimum solution. Thus new ideas usually emerge for later consideration.
3. The group should be open minded.

   All of the participants should be mentally alert individuals. They should not be hierarchically structured. And, if possible, they should be made up of equals, to prevent the censoring of such thoughts as might give offense to superiors or subordinates. Usually people who habitually emphasize negative aspects should not be invited, and executives or other people mostly concerned with evaluation and judgment do not make good participants.

Reporter

One or two participants should be appointed by the leader to keep a stenographic account of the ideas offered by the group during the brainstorming period. A tape recorder is usually used to help in this.

5.4.4 Brainstorming Session

In general, the entire brainstorming session should not last more than thirty minutes to an hour, or should be wound up when no more new ideas are forthcoming. Experience has shown that longer sessions produce nothing new and lead to unnecessary repetitions. However, group members may add ideas to the accumulated list of ideas for a 24-hour period.

During a brainstorming session, the group leader explains the problem statement first, then the group members express their ideas, and the reporter records all the ideas.

Problem statement

The group leader should make every effort to describe the problem in clear and concise terms. Some typical examples of ideas that might satisfy the problem statement may be included with the statement.

Particular care should be taken to confine the problem statement within a narrow range to ensure that all group members direct their ideas toward a common target. If the problem is stated too narrowly, then the range of ideas from the session may be rather limited. On the other hand, the problem should be specific, rather than general, in nature. A very vague problem statement leads to equally vague ideas that may be impractical.

The group leader first states the design problem, including known requirements. A written statement should be provided for each participant if possible. Questions and discussions are carried on until the participants understand the problem. The problem statement may be modified.

Idea brainstorming

In response to the initial problem statement, the group members are asked to spend a few minutes in silence to write down the first ideas that come into their minds. The introduction of a preliminary period for writing ideas down is a neat way of avoiding the risk of delays or failures when group members have not learnt to trust each other enough to speak everything on their minds.

It is a good idea if each member has small record cards on which to write ideas. The ideas should be expressed succinctly, written one per card, and solid enough to allow the emergence of specific solution ideas. The next and major part of the session is for each member of the group, in turn, to read out one idea from his set. Having ideas on cards greatly reduces the time needed to classify the results.

Suggested ideas are then listed in short form on a blackboard, by the group leader and one or more assistants, for all to see.

Also, the checklist method for providing generalized questions has proven helpful during brainstorming sessions.

Idea recording

A group brainstorm session usually produces a long list of potential idea solutions, usually five to nine per participant in a session. They should be carefully documented and summarized in preparation for evaluation. Recording ideas on a blackboard is a good approach. A tape recorder can be very valuable, especially when participants quickly suggest several different ideas.

5.4.5 Brainstorming Rules

In order to have an effective brainstorming session, the following essential rules must be agreed upon by the group before the session.

Criticism is not allowed

The most important rule of a brainstorming session is that no criticism is allowed from any other member of the group during the session. Participants must understand that the objective of the session is to create a supportive environment for free flowing ideas, and they must not become involved in personal quarrels.

It is an interesting characteristic of brainstorming session that some ideas that seem ridiculous at first may later turn out to be extremely useful for further considerations. For this reason, appraisals, rejection, analysis, judgments, criticisms, or ridicule in any form must be withheld until after the brainstorming session.

The usual responses to unconventional ideas, such as "It's silly", "It cannot be done", "It will never work", "It has nothing to do with the problem", or "We've heard it all before", usually scare off spontaneity and creativity.

Free wheeling is expected

A brainstorming session should be fun, and should be conducted in a free and informal manner. The atmosphere should be fully relaxed and free wheeling.

In brainstorming sessions, all ideas are encouraged. Participants should speak out all ideas entering their minds without any constraint. They should avoid rejecting any absurd, false, embarrassing, stupid, or redundant ideas expressed spontaneously by themselves or by other members of the group. Many ideas that are normally held back because of fear of ridicule and criticism are now presented. The wilder the idea, the better. Furthermore, the practicability of the suggestions should be ignored at first.

Quantity is wanted

The brainstorming group should be encouraged to think up as many ideas as possible. Experience has shown that the larger the number of generated ideas, the higher the probability of finding an outstanding solution. In many instances, ideas that would normally have been omitted turn out to be the best ideas.

It is not unusual for a group of six to generate 30 to 50 ideas in a half hour of brainstorming. Obviously, to achieve such a quantity of output the ideas can only be described roughly and without details.

Combination and improvement are sought

In brainstorming sessions, the ideas generated by a member may serve as stimulus to another, producing a chain reaction. Someone else may suggest a modification that improves the feasibility of an early concept, or even make a previously impractical idea successful. The result may be the generation of entirely new solutions that combine the input of several group members. Brainstorming, then, provides an chance to capitalize on the ideas of others, compounding the benefits derived from the response to each idea presented.

5.4.6 Brainstorming Evaluation

The total output of ideas should be classified into related groups under logical headings. Normally, the major headings are the basic design concepts. Then, the entire list of ideas should be rigorously evaluated, analyzed, criticized, eliminated, and discussed, either by the original brainstorming group or preferably by a new team.

Individual evaluation

The developed list of design concepts should be sent to each participant. The group members then work individually to identify designs that seem best in meeting design specifications. Each member should provide three to five solutions, in addition to the evaluation criteria used.

Each member then reviews all proposed solutions and selects one approach that seems best in meeting all requirements, and another as first alternate. In addition to explaining the design concepts in words, each member is encouraged to make a decision table and rough sketches of chosen design concepts.

Although this evaluation process may be imperfect, conceptual ideas must be evaluated subjectively until the number of potential solutions is narrowed sufficiently to allow a thorough objective evaluation within the scope of available resources. The leader and assistants then write out chosen selections and the list of evaluation criteria in preparation for reconvening the whole group.

Group evaluation

The group meets for a final review of all design concepts to avoid possible misunderstandings or one sided interpretations on the part of the experts.

The group is reconvened for the purpose of improving the evaluation criteria and recommended solutions. First, the list of evaluation criteria is reviewed and refined by the group. Then, each participant presents potential solutions and the group provides ideas for improvement in a second brainstorm. The group leader then summarizes the evaluation criteria and all the recommended design concepts in their improved form on the blackboard. New and more advanced ideas may well be expressed or developed during such a review session.

5.4.7 Brainstorming Report

The group leader and assistant then integrate the individual and group evaluations into one comprehensive decision table that rationalizes the several recommended solutions. This, together with the individual sketches, constitutes the final report.

A copy of the final report should be given to each participant for his further comments. The group leader collects these reports after review and uses them in the final judgments of the design concept selected.

5.4.8 Examples

Two examples are provided to illustrate how the free flow of ideas is stimulated by brainstorming sessions.

[Example 5.5]
Lady bicycles.

This is a classroom example conducted by the lecturer of the class to demonstrate the brainstorm technique.

Six out of fifty students in the class, selected randomly by the lecturer, participate in the group session. They are all senior students in mechanical engineering. The problem is to generate ideas for suggesting extra functions and performance of lady bicycles. Two other students are assigned by the group leader, i.e., the lecturer, as the reporters. A blackboard is used to document the ideas.

The group leader orally addresses the problem first. Since the problem itself is quite clear, a written problem statement is not necessary in this case. Then, the six students are asked to spend three minutes in silence to write down the immediate ideas that come to their minds – on small record cards, one idea per card.

The brainstorming session lasts for about 35 minutes, and fifty ideas are recorded. These ideas are grouped under nine major categories, as follows:

1. Power source
   Power by leg(s), ankle(s), hand(s), finger(s), body motions.
2. Drive train
   Belt drive, gear drive, cable drive, linkage drive, hybrid drive.
3. Transmission
   Manual transmission, automatic transmission, continuous variable transmission, discrete speed transmission, hybrid transmission.
4. Steering
   Steering by head, steering by sight, steering by voice, steering handle with adjustable angles, steering handle with adjustable heights, adjustable casting angles, wheel type steering handle.
5. Frame
   Uni-body, portable, foldable, shrinkable, front and/or rear bumpers.
6. Brake
   Disk brake, air aided brake, anti-lock brake system, body brake.
7. Shock absorption
   Seat suspension, rear wheel suspension, front wheel suspension, handle bar shock absorption, air suspension, anti-dive suspension.
8. Safety
   Air bag, seat belt, signal lights, rear view mirror, built-in lock, push bottom automatic lock.
9. Others
   Fan, ash tray, leather seat, constant temperature seat, beverage holder, umbrella holder, convertible top, foldable windshield, radio, tape, compact disc, auxiliary wheels for balancing and standing.

[Example 5.6]
Mechanical animals.

The vice-president of the R&D division of a toy company, after some feasibility study, has decided to develop a series of mechanical animals for the coming Christmas market that is six months away. A design engineer is appointed as the project manager for this task.

The design engineer starts with the preliminary design of this project. Although the problem delivered by the vice-president is simple and clear, it is not well defined. Quantitative and even qualitative specifications regarding mechanical animals are not mentioned. The project leader realizes that he needs to come up the detailed problem definition by himself and to generate fresh ideas to guarantee a successful market. He decides to have a brainstorming session for this purpose.

Group

Nine people, including the design engineer, participate in the session. Variety of experience is sought to broaden the idea output. Group members consist of the project manager as the leader, one research engineer, one associate professor in biology, one retired lawyer, one writer, one secretary, one bartender, one high school student, and one ten year old boy. The secretary is assigned as the reporter.

Session

The group leader orally explains the problem, with the aid of a written statement on a transparency. Two flip charts and a whiteboard are used to document the ideas. The session continues for about 55 minutes. Generated ideas are recorded and grouped under the following seven design headings:

1. Size
   Pocket size, football size, real animal size.
2. Type of animals
   Rat, ox, tiger, rabbit, dragon, snake, horse, goat, monkey, chicken, dog, pig.
3. Input power
   Gravity, manpower, spring, balloon air, battery, shaking motion, noise energy.
4. Gait
   Walk, pace, trot, canter, gallop, bound.
5. Function
   Forward and backward motion, turning motion with signal lights, swimming with tail as the pedal, flying, with or without remote control, singing during motion.
6. Mechanism
   Linkages (4-bar, 6-bar, 8-bar, 10-bar), linkages and cables or wires, linkages and gears.
7. Material
   Plastic, paper, acrylic, wood, brass, bronze, steel.

Evaluation

In the following week, the project manager conducts a one day off-site evaluation meeting. The ideas generated in the brainstorming session are critically reviewed by an evaluation team, consisting of the project manager as the group leader, the research engineer in the brainstorming session, one development engineer, one marketing representative, one industrial designer, one shop floor technician, one manufacturing engineer, one artist, and one materials specialist. The performance specifications and other business requirements are critically reviewed.

On this basis, the project manager decides to design a series of high quality mechanical animals that are versatile in function, reliable in usage, and graceful in outlook. The following concepts are finally adapted for the detailed design:

1. The product is defined as a precious gift rather than a popular toy.
2. The target customers are company VIPs rather than teenagers.
3. The product should be a limited edition rather than mass production.
4. The size of the product should be small than a football for VIPs' tables or cabinets.
5. The gait of the machine animals is "walk" as the first year's target.
6. The type of animals is limited to those that have the same gait of walk as horses, for examples, oxen, tigers, dogs, ... etc., so as to use the same drive mechanism.
7. The mechanical animals can walk forwards and backwards and have built-in chips to sing ten songs.
8. The legs of the mechanical animals are driven by linkages.
9. The material for the animal body is fine wood with treasure stones as optional decorations. The outlook of the animal body can be custom made by contracted artists or even provided by the customer.

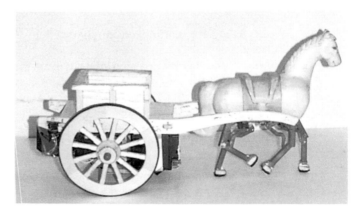

Figure 5.6 A walking mechanical horse (see Appendix)

Figure 5.6 shows a mechanical horse that is powered by a motor with battery under the added on trailer and is remote controlled. This horse walks forwards and backwards. Each leg is driven by an 8-bar linkage. The body of the horse is wood that is hand carved by a craftsman.

5.5 Summary

A matrix is an array of elements that can be manipulated according to certain rules. *Matrix technique* is a creative method of providing a large variety of design concepts starting from a few basic ideas for major product functions. There are several types of matrix approaches available to help generate creativity. However, attribute listing and morphological chart analysis are the most commonly used.

The creative technique of attribute listing is used as a group effort to identify the essential attributes of the problem and to stimulate the desired solution.

Morphological chart analysis, an individual creative effort, provides a useful means of obtaining a wide variety of possible design concepts. Different combinations of subsolutions can be selected from the chart, perhaps pointing to new solutions.

Brainstorming is an "everything goes" problem solving conference method, use on a subject from many different perspectives in an open environment. It usually produces a large number of alternative ideas. However, most of the generated ideas will subsequently be discarded, with perhaps a few novel ideas being identified as worth following up.

Different creative techniques should be used in combination so as to best meet particular design problems. Furthermore, creative techniques and rational methods are complementary aspects of a systematic approach to design. Design engineers should understand them thoroughly, to best use them in generating ideas.

Problems

5.1 List the attributes of the following designs:
   (a) desk lamps.
   (b) computer tables.
   (c) mountain bikes.
5.2 Suggest possible improvements, based on the technique of attribute listing, for the following designs:
   (a) sofa beds.
   (b) bicycle stands.
   (c) motorcycles.
5.3 Develop morphological charts for the following designs:
   (a) door locks.
   (b) rocking chairs.
   (c) mini cars.
5.4 Study your strategy of job hunting based on the technique of morphological chart analysis.

5.5 Conduct a brainstorming session to suggest possible uses for the following items:
(a) used newspapers.
(b) used plastic beverage bottles.
(c) used cars.

5.6 Conduct an individual brainstorming session to come up with ideas to improve wheelchairs.

5.7 Repeat Problem 5.6 based on group brainstorming.

References

Alger, J. R. M. and Hays, C. V., Creative Synthesis in Design, Prentice-Hall, 1964.

Asimow, M., Introduction to Design, Prentice-Hall, 1962.

Beakley, G. C. and Leach, H. W., Engineering - An Introduction to A Creative Profession, Macmillan, 1967.

Chiou, C. P., On the Design of A Wave Gait Walking Horse, M.S. thesis, Department of Mechanical Engineering, National Cheng Kung University, Tainan, Taiwan, June 1996.

Cross, N., Engineering Design Methods, John Wiley & Sons, 1994.

Dieter, G. E., Engineering Design, McGraw-Hill, 1983.

Edel, Jr., D. H., Introduction to Creative Design, Prentice Hall, 1967.

Hill, P. H., The Science of Engineering Design, Holt, Rinehart and Winston, 1970.

Jones, J. C., Design Methods, John Wiley & Sons, 1980.

Knoblock, E. W. and Ong, J. N., Introduction to Design, Spectra, 1977.

Lewis, W. and Samuel, A., Fundamentals of Engineering Design, Prentice Hall, 1989.

Middendorf, W. H., Engineering Design, 1969.

Pahl, G. and Beitz, W., Engineering Design, Springer-Verlag, 1977.

Shoup, T. E., Fletcher, L. S., and Mochel, E. V., Introduction to Engineering Design, Prentice-Hall, 1981.

Vidosic, J. P., Elements of Design Engineering, Ronald Press, Allyn and Bacon, 1969.

Wang, Y. C., Conceptual Design of Sequential and Repeatable Push-button Locks with Variable Passwords, M.S. thesis, Department of Mechanical Engineering, National Cheng Kung University, Tainan, Taiwan, June 1996.

Wright, P. H., Koblasz, A., and Sayle, W. E., Introduction to Engineering, John Wiley & Sons, 1989.

A Creative Design Methodology

CHAPTER 6

CREATIVE DESIGN METHODOLOGY

M*ethodology* is a set of procedures or design steps for solving a defined problem systematically. It includes a body of methods, rules, and/or postulates. This chapter presents a creative design methodology, based on modifications of existing devices, for the generation of all possible topological structures of mechanical devices. An automotive front suspension mechanism is adopted as an example. Detailed descriptions of major design steps and the applications of this design methodology are treated in detail in later chapters (Chapters 7-14).

6.1 Introduction

Experience is the knowledge, skill, or practice derived from direct observation of, or participation in, events. It provides the richest knowledge that one may recall whenever needed. Therefore, when an engineer faces a target project, experience is the best approach for producing design concepts.

An inexperienced engineer may start to solve the problem by traditional rational methods, such as analysis of existing designs, information search, and checklist, as presented in Chapter 4. And he should gain experience through reading, listening, speculating, observing, tinkering with devices, and investigating the multitude of products to find out how they work. Furthermore, the design engineer can apply creative techniques, such as attribute listing, morphological chart analysis, and brainstorming, as presented in Chapter 5, to help generate ideas in conceptual design.

Although design concepts can be brought into existence with the aid of rational methods and creative techniques serving as a catalyst for the concept derivation, these organized methods, nevertheless, are not quite precise. So far, no existing method is available to guide engineering designers directly to invent products. However, to help achieve this goal, this chapter presents a step by step creative design methodology for the systematic generation of all possible topological structures of mechanical devices that perform the same or similar tasks as available existing designs, and subject to certain design requirements and constraints.

6.2 Procedure

Figure 6.1 shows the flowchart of the proposed methodology for the creative design of mechanical devices. The steps are:

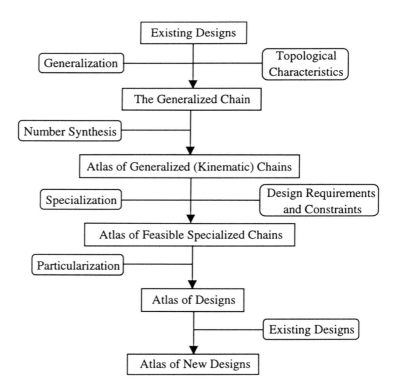

Figure 6.1 The creative design methodology

Step 1. Identify existing designs with required design specifications that designers would like to have, and conclude the topological characteristics of these designs.

Step 2. Select an existing design arbitrarily, and transform it into its corresponding generalized chain, according to the rules of generalization described in Chapter 7.

Step 3. Synthesize the atlas of generalized chains that have the same numbers of members and joints as the generalized chain obtained in Step 2, based on the algorithm of number synthesis presented in Chapter 8 or Chapter 9, or simply select the needed atlas from available atlases of chains in these two chapters.

Step 4. Assign types of members and joints to each generalized chain obtained in Step 3, to have the atlas of feasible specialized chains based on the algorithm of specialization presented in Chapter 10, to meet needed design requirements and constraints.

Step 5. Particularize each feasible specialized chain obtained in Step 4 into its corresponding schematic format of mechanical device, to have the atlas of mechanical devices.

Step 6. Identify existing designs from the atlas of designs, to have the atlas of new designs.

6.3 Existing Designs

The first step of the creative design methodology is to define the design specifications of mechanical devices that design engineers would like to generate. Then search and study available existing designs with the required specifications, and conclude the fundamental topological characteristics of these designs.

Specification is a detailed precise written statement describing a product. It should be defined in the beginning of the design process of any mechanical device. Without a clear statement of product specifications, the design process of the product cannot be rigidly carried out. Furthermore, specification is product oriented. Different products, or products with different performances, have different specifications. And for engineering problems, any solution that does not meet the specifications is worthless.

In the beginning of the conceptual phase for creating mechanical devices, only basic specifications regarding topological structures of the designs are of major concern. Those specifications relating to the products' motion range, force scale, work capacity, efficiency, performance, ... etc., can be neglected at this stage.

In what follows, an automotive front suspension mechanism is used as an example.

An automobile is designed to provide ground transportation, with the ground not necessarily solid and flat. Therefore, a suspension system is needed to absorb road shocks and to provide a reasonably comfortable ride for passengers. The springs in the car seats absorb some shocks from the car body to the passengers. The tires absorb some of the irregularities in the road to the car body. Suspension mechanisms between the tires and the car body are further needed to absorb most of the shock from the road to the car body. Various types of suspension mechanisms have been designed for this purpose, such as coil spring suspensions, MacPherson front suspensions, and torsion-bar front suspensions, just to name a few.

Suppose a suspension engineer would like to create all possible design configurations of automotive front suspension mechanisms such that each one has the following basic design specifications:
1. It is an independent front suspension.
2. It has one independent input.
3. It is a spatial five-bar mechanism.

And suppose the engineer finds that the design shown in Figure 6.2 meets the required specifications. Then, the characteristics of the topological structure of this design are concluded as follows:

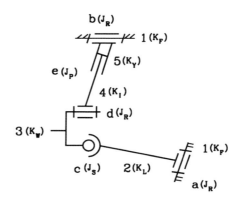

Figure 6.2 **An existing design of automotive front suspension mechanisms**

1. It consists of five members and five joints.
2. It has one ground link (KF, member 1), one kinematic link (KL, member 2), one wheel link (KW, member 3), and one shock absorber consisting of a piston (KI, member 4) and a cylinder (KY, member 5).
3. It has three revolute joints (JR, joints a, b, and d), one prismatic joint (JP, joint e), and one spherical joint (JS, joint c).

4. It is a spatial mechanism with one degree of freedom. The topology matrix, M_T, of this mechanism is:

$$M_T = \begin{bmatrix} K_F & J_R & 0 & 0 & J_R \\ a & K_L & J_S & 0 & 0 \\ 0 & c & K_W & J_R & 0 \\ 0 & 0 & d & K_I & J_P \\ b & 0 & 0 & e & K_Y \end{bmatrix}$$

6.4 Generalization

The second step of the creative design methodology is to select a design from available existing designs to serve as an original design to continue the design process. Any existing design can be selected as the original design. This original design is then transformed into its corresponding generalized (kinematic) chain.

The purpose of generalization is to transform the original mechanical device, that involves various types of members and joints, into a generalized chain with only generalized links and (revolute) joints. The process of generalization is based on a set of generalizing rules that are derived according to defined generalizing principles. These generalized principles and rules will be described in detail in Chapter 7.

For the existing design shown in Figure 6.2, its corresponding generalized chain, consisting of five generalized links and five generalized joints, is shown in Figure 6.3.

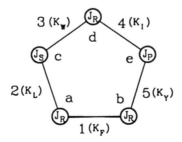

Figure 6.3 The generalized chain and the atlas of generalized chain of the automotive front suspension mechanism

Through the process of generalization, designers are able to study and compare different designs in a very basic way. Mechanical devices that at first glance appear to be different may have identical generalized forms.

6.5 Number Synthesis

The third step of the creative design methodology is to synthesize all possible generalized chains that have the same numbers of links and joints as the original generalized chain. This is kinematic number synthesis and has been a subject of numerous studies in past years.

Chapters 8 and 9 will present detailed algorithms for obtaining atlases of generalized chains and kinematic chains, respectively, with the required numbers of links and joints. As a design engineer in industry, one can simply select the needed atlas from the results of these two chapters and skip the complicated algorithms.

In the example, there is only one generalized chain with five links and five joints, i.e., the one shown in Figure 6.3.

6.6 Specialization

The fourth step of the creative design methodology is to assign specific types of members and joints to each available generalized chain, subject to certain design requirements to have specialized chains. Design requirements are determined based on the concluded topological structures of existing designs.

In the example of an automotive front suspension mechanism, the design requirements are as follows:
1. There must be a ground link (KF) as the car body.
2. There must be a shock absorber, consisting of a cylinder (KY) and a piston (KI) and adjacent to both the ground link and the wheel link, to absorb road shocks.
3. There must be a wheel link (KW), not adjacent to the ground link, to install the wheel.
4. There are three revolute pairs, one prismatic pair, and one spherical pair. And the joint incident to the cylinder and the piston must be a prismatic pair.

Figure 6.4 shows the resulting atlas of specialized chains obtained from the generalized chain shown in Figure 6.3, subject to the above design requirements.

Feasible specialized chains are identified from the available atlas of specialized chains, subject to certain design constraints. Design constraints are defined based on engineering reality and designers' decisions. These constraints can be flexible, and they are varied for different cases.

In the example of an automotive front suspension mechanism, let the design constraints be:
1. The cylinder (KY) and the piston (KI) should not be taken as the ground link (KF).

2. The spherical joint should not be incident to the ground link.

Figures 6.4 (a) and (b) are two feasible specialized chains subject to these design constraints. If the constraint that the spherical joint should not be incident to the ground link is released, four feasible specialized chains are available, as shown in Figures 6.4 (a)-(d). And if the constraint that the cylinder and the piston should not be taken as the ground link is further released, all eight specialized chains shown in Figure 6.4 are feasible.

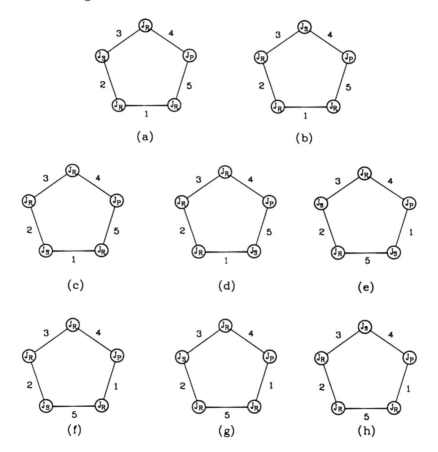

Figure 6.4 Atlas of (feasible) specialized chains of the automotive front suspension mechanism

Chapter 10 will present an algorithm for the enumeration and counting of all possible non-isomorphic specialized chains, based on combinatorial theory and Polya's theory.

6.7 Particularization

Once a feasible specialized chain is obtained, it is particularized into its corresponding mechanical device in a skeleton drawing.

Graphically, *particularization* is the reverse process of generalization, and can be done by applying the generalizing rules in reverse order. Figure 6.5 shows the corresponding atlas of designs for the atlas of (feasible) specialized chains shown in Figure 6.4.

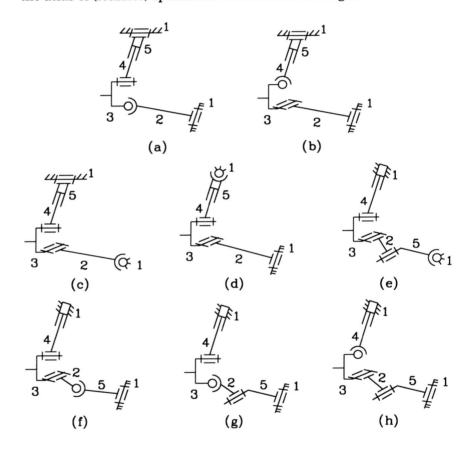

Figure 6.5 Atlas of designs for the automotive front suspension mechanism

6.8 Atlas of New Designs

The last step of the creative design methodology is to identify all existing designs from the atlas of designs. Then, those that are not identified as existing designs are new designs.

Existing designs are normally identified from commercial products and exhaustive patent search. Each identified new design gives design engineers an opportunity to avoid patent protections of existing designs and may even lead to a possible new patent.

The design shown in Figure 6.5(a) is the existing and original design. Therefore, the other seven designs shown in Figure 6.5 are new designs for the topological structures of the mechanical devices.

Problems

6.1 Provide one existing and patented design of mechanisms, describe its function, list its specifications, identify its topology matrix, and conclude its topological characteristics.

6.2 For the existing mechanism described in Problem 6.1, conclude its design requirements.

6.3 For the existing mechanism described in Problem 6.1, discuss its design constraints.

6.4 For the existing mechanism described in Problem 6.1, provide a new concept to avoid patent protection.

6.5 Provide one existing and patented design of clamping devices, describe its function, list its specifications, identify its topology matrix, and conclude its topological characteristics.

6.6 For the existing clamping device described in Problem 6.5, conclude its design requirements.

6.7 For the existing clamping device described in Problem 6.5, discuss its design constraints.

6.8 For the existing clamping device described in Problem 6.5, provide a new concept to avoid patent protection.

6.9 Synthesize all possible design configurations of automotive front suspension mechanisms, based on the existing design shown in Figure 6.2, without any design constraint.

References

Chen, J. J., Type Synthesis of Planar Mechanisms, Master thesis, National Cheng Kung University, Tainan, Taiwan, June 1982.

Hwang, Y. W., An Expert System for Creative Mechanism Design, Ph.D. dissertation, Department of Mechanical Engineering, National Cheng Kung University, Tainan, Taiwan, May 1990.

Yan, H. S., "A methodology for creative mechanism design," *Mechanism and Machine Theory*, Vol. 27, No. 3, 1992, pp. 235-242.

Yan, H. S. and Chen, J. J., "Creative design of a wheel damping mechanism," Mechanism and Machine Theory, Vol. 20, No. 6, 1985, pp. 597-600.

Yan, H. S. and Hsieh, L. C., "Concept design of planetary gear trains for infinitely variable transmissions," Proceedings of 1989 International Conference on Engineering Design, Harrogate, England, August 22-25, 1989, pp.757-766.

Yan, H. S. and Hsu, C. H., "A method for the type synthesis of new mechanisms," *Journal of the Chinese Society of Mechanical Engineers (Taiwan)*, Vol. 4, No. 1, 1983, pp. 11-23.

CHAPTER 7

GENERALIZATION

Mechanical devices involving various types of elements can be transformed into corresponding generalized (kinematic) chains with only generalized (revolute) joints and generalized (kinematic) links. Through generalization, design engineers are able to study and compare different devices in a very basic way. Mechanical devices that at first glance appear to be quite different may have identical generalized forms.

Generalization is one of the major steps of the creative design methodology for generating mechanical devices. The concept of generalization is based on a set of generalizing rules. These generalizing rules are derived according to defined generalizing principles. The purpose of this chapter is to develop the rules and conventions for the generalization. The definitions of generalized joints and generalized links are established first. Then generalizing principles and generalizing rules are defined, and the process for obtaining generalized (kinematic) chains is presented. Finally, several examples for the generalization are given. Figure 7.1 shows the flowchart of the process of generalization.

7.1 Generalized Joints and Links

A *generalized joint* is a joint in general; it can be a revolute pair, prismatic pair, spherical pair, helical pair, or others. A generalized joint with two incident members is called a *simple generalized joint*, whereas a generalized joint with more than two incident members is called a *multiple generalized joint*. A simple generalized joint with N_L

incident members is graphically symbolized by N_{L-1} small concentric circles with a dot in the center. Figures 7.2(a), (b), and (c) show a generalized joint with two, three, and four incident members, respectively. Besides, Figure 7.2(a) is a simple generalized joint, and Figures 7.2(b) and (c) are multiple generalized joints.

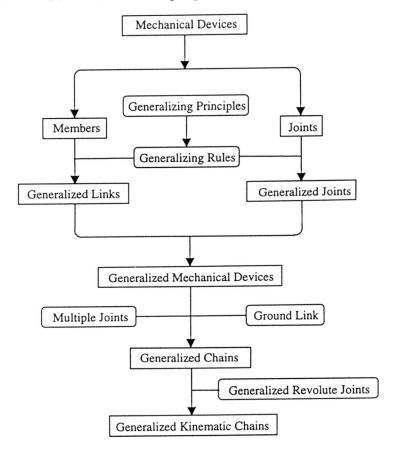

Figure 7.1   The process of generalization

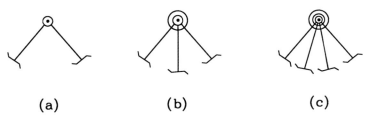

Figure 7.2   Representations of generalized joints

A *generalized link* is a link with generalized joints; it can be a binary link, ternary link, quaternary link, ... etc. Graphically, a generalized link with N_J incident joints is symbolized by an N_J-sided, crosshatched polygon with small circles that have a dot in the center as vertices. Figures 7.3(a), (b), and (c) show a binary, ternary, and quaternary generalized link, respectively.

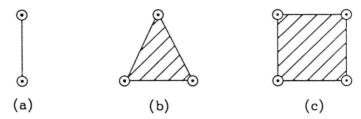

(a)   (b)   (c)

Figure 7.3 **Representations of generalized links**

7.2 Generalizing Principles

The fundamental strategy for transforming the schematic diagram of a mechanical device into its corresponding generalized (kinematic) chain is based on the following generalizing principles:

1. All joints between members of mechanical devices are transformed into generalized (revolute) joints.
2. All members of mechanical devices are transformed into generalized links.
3. The topological incidency and adjacency among members and joints of a mechanical device and its corresponding generalized (kinematic) chain should be the same.
4. The number of degrees of freedom for the mechanical device and its corresponding generalized (kinematic) chain should be the same.

These generalizing principles are the basic laws for developing rules for the process of generalization. Without these, the process of creative design would be wide open, and would therefore not be systematic and precise.

7.3 Generalizing Rules

In the following, a set of rules developed, based on the above generalizing principles, for the process of generalization of mechanical devices is presented.

Joints
a. A revolute pair is replaced by a generalized joint with letter J_R indicating the revolute pair, Figure 7.4(a).
b. A prismatic pair is replaced by a generalized joint with letter J_P indicating the prismatic pair, Figure 7.4(b).
c. A rolling pair is replaced by a generalized joint with letter J_O indicating the rolling pair, Figure 7.4(c).
d. A cam pair is replaced by a generalized joint with letter J_A indicating the cam pair, Figure 7.4(d).
e. A gear pair is replaced by a generalized joint with letter J_G indicating the gear pair, Figure 7.4(e).
f. A wrapping pair is replaced by a generalized joint with letter J_W indicating the wrapping pair, Figure 7.4(f).
g. A helical pair is replaced by a generalized joint with letter J_H indicating the helical pair, Figure 7.4(g).
h. A cylindrical pair is replaced by a generalized joint with letter J_C indicating the cylindrical pair, Figure 7.4(h).
i. A spherical pair is replaced by a generalized joint with letter J_S indicating the spherical pair, Figure 7.4(i).
j. A flat pair is replaced by a generalized joint with letter J_F indicating the flat pair, Figure 7.4(j).
k. A universal joint is replaced by a generalized joint with letter J_U indicating the universal joint, Figure 7.4(k).
l. A direct contact is replaced by a generalized joint with letter J_D indicating the direct contact, Figure 7.4(l).

Members
a. A link with N_L adjacent members is replaced by a generalized link with N_L generalized joints, Figure 7.5(a).
b. A slider with N_L adjacent members is replaced by a generalized link with N_L generalized joints, Figure 7.5(b).
c. A roller with N_L adjacent members is replaced by a generalized link with N_L generalized joints, Figure 7.5(c).
d. A cam with N_L followers is replaced by a generalized link with N_{L+1} generalized joints, Figure 7.5(d).
e. A gear with N_L adjacent gears is replaced by a generalized link with N_{L+1} generalized joints, Figure 7.5(e).
f. A pulley (or sprocket) with a belt or rope (or chain) wrapping on it is replaced by a ternary generalized link, Figure 7.5(f).
g. That portion of a belt or rope (or chain) not in contact with the pulley (or sprocket) is replaced by a generalized binary link; and that portion of a belt or rope (or chain) in contact with the pulley (or sprocket) is neglected, Figure 7.5(g).

7.3. Generalizing Rules • 101

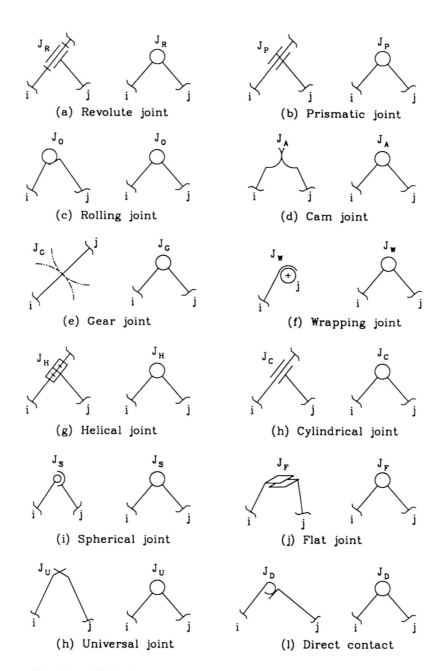

Figure 7.4 Generalization of joints

h. An actuator, consisting of a piston and a cylinder, is replaced by a pair of generalized binary links incident to a generalized prismatic pair, Figure 7.5(h).
i. A spring is replaced by a generalized binary link if the mechanical device functions as a structure, Figure 7.5(i_1); and is replaced by a pair of generalized binary links incident to a generalized revolute joint if the mechanical device functions as a mechanism, Figure 7.5(i_2).
j. An applied force is replaced by a generalized binary link with a generalized revolute joint for planar mechanical devices, Figure 7.5(j_1); and with a generalized spherical joint for spatial mechanical devices, Figure 7.5(j_2).

This set of generalizing rules is not the only set conceivable, nor is it a closed set. For those joints and members not mentioned here, additional rules could be defined according to the defined generalizing principles.

For some special purposes that will be clear later, all generalized joints with specific types of kinematic pairs can further be transformed into *generalized revolute joints* or generalized links with generalized revolute joints only. In the following, the rules for the transformation of generalized joints are described:

a. A generalized prismatic joint is replaced by a generalized revolute joint, Figure 7.6(a).
b. A generalized rolling joint is replaced by a generalized revolute joint, Figure 7.6(b).
c. A generalized cam joint is replaced by a binary link with a generalized revolute joint at both ends, Figure 7.6(c).
d. A generalized gear joint is replaced by a binary link with a generalized revolute joint at both ends, Figure 7.6(d).
e. A generalized wrapping joint is replaced by a generalized revolute joint, Figure 7.6(e).
f. A generalized helical joint is replaced by a generalized revolute joint, Figure 7.6(f).
g. A generalized cylindrical joint is replaced by a binary link with a generalized revolute joint at both ends, Figure 7.6(g).
h. A generalized spherical joint is replaced by a dyad with incident generalized revolute joints, Figure 7.5(h).
i. A generalized flat joint is replaced by a dyad with incident generalized revolute joints, Figure 7.6(i).
j. A universal joint is replaced by a dyad with incident generalized revolute joints, Figure 7.6(j).
k. A direct contact is replaced by a binary link with a generalized revolute joint at both ends for planar structures, and with a generalized spherical joint at both ends for spatial mechanisms, Figure 7.6(k).

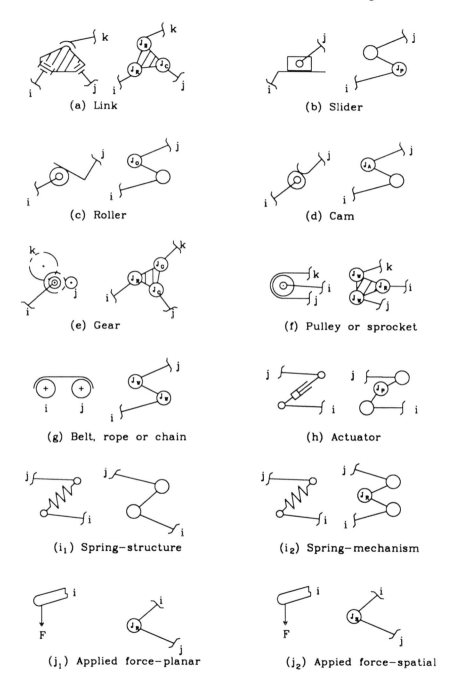

Figure 7.5 Generalization of members

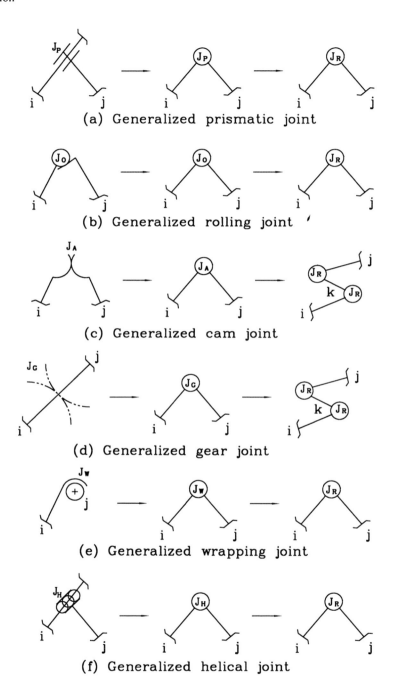

Figure 7.6 Generalization of generalized joints

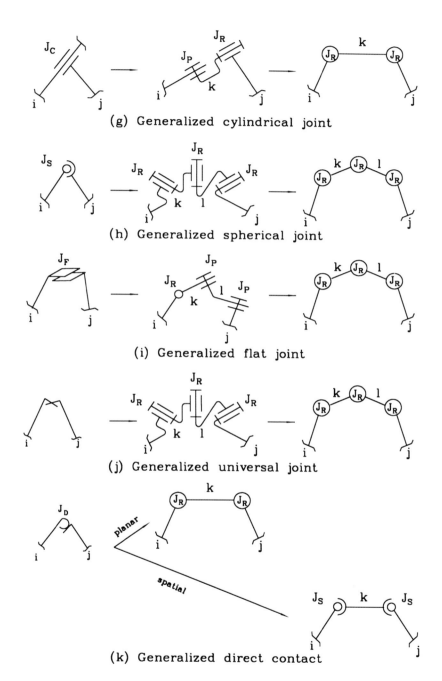

Figure 7.6 Generalization of generalized joints (continued)

The rules for the transformation of any other type of joints can also be defined in the same way. The point is that the result after generalization should have the same degrees of constraint as the original generalized joint.

7.4 Generalized (Kinematic) Chains

A *generalized mechanical device* is obtained by applying generalizing rules to the schematic diagram of a mechanical device. A *generalized chain* is obtained by releasing the ground link and eliminating multiple joints of the corresponding generalized mechanical device. If a generalized chain is further transformed into one with only generalized revolute joints, it is a *generalized kinematic chain*.

The elimination of a multiple joint can be carried out by increasing the number of joints of any incident member of this joint, one at a time, until this original joint has only two incident members. However, if a multiple joint is incident to the ground, the ground link should be the only member to be expanded. Figure 7.7(a) shows a multiple joint with four incident binary links (links i, j, k, and l). By transforming any one of these binary links into a ternary link, a multiple joint with three incident links is obtained. Figures 7.7(b_1)-(b_{12}) show 12 solutions of this transformation. Then, by increasing the number of joints of any one of those three incident links of the multiple link shown in Figures 7.7(b_1)-(b_{12}) by one, a solution without any multiple joint is obtained. Figures 7.7(c_1)-(c_3) show three solutions obtained from the one shown in Figure 7.7(b_1).

Based on Cayler's theorem of graph theory for counting labeled trees, the number of possible ways to eliminate a multiple joint with N_{Li} (>3) incident links is $N_{Li}^{N_{Li}-2}$. For the multiple joint shown in Figure 7.7(a), $N_{Li}=4$, there are $4^{4-2}=16$ non-isomorphic solutions with only simple joints, as shown in Figure 7.7(d_1)-(d_{16}).

Figure 7.8 shows a simple example for transforming a spring-linkage mechanism with a multiple joint into its corresponding generalized chains. All joints in this device are revolute pairs. First, the spring (K_S) shown in Figure 7.8(a) is replaced by a pair of generalized binary links (links 5 and 6) with an incident generalized revolute joint f as shown in Figure 7.8(b). Then, the ternary ground link (link 1) is released as shown in Figure 7.8(c). Since $N_{Li}=3$, there are $3^{3-2}=3$ possible ways to eliminate multiple joint e by adding one more simple joint g into links 3, 4, and 5, as shown in Figures 7.8(d_1), (d_2), and (d_3), respectively. Those three chains shown in Figures 7.8(d_1), (d_2), and (d_3) are the corresponding generalized chains. The two chains shown in Figures 7.8(d_1) and (d_3) are isomorphic. Since all joints are revolute joints, there are also generalized kinematic chains.

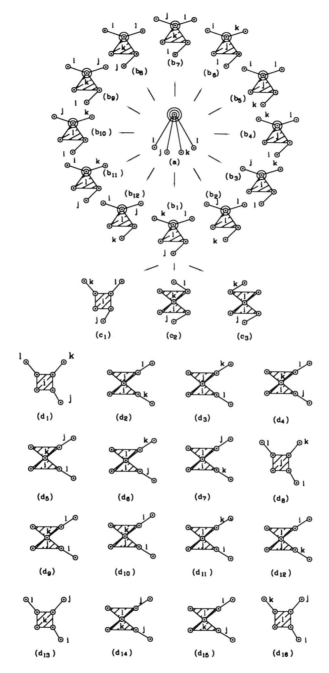

Figure 7.7 Elimination of a multiple joint

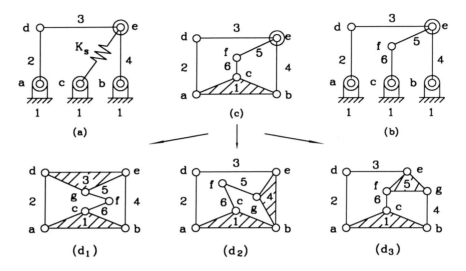

Figure 7.8 Generalization of a spring-linkage mechanism

7.5 Examples

Several examples are given as follows to illustrate the process of generalization.

[Example 7.1]
A slider-pulley-spring mechanism, Figure 7.9(a).
 This is a (5,6) mechanism. The five members are: the ground link (member 1, K_F), the slider (member 2, K_P), the belt (member 3, K_B), the pulley (member 4, K_U), and the spring (member 5', K_S). And the six joints are: joint a (members 1 and 2; J_P), joint b (members 1 and 4; J_R), joint c (members 1 and 5'; J_R), joint d (members 2 and 3; J_R), joint e (members 3 and 4; J_W), and joint f (members 4 and 5'; J_R). Its topology matrix, M_T, is:

$$M_T = \begin{bmatrix} K_F & J_P & 0 & J_R & J_R \\ a & K_S & J_R & 0 & 0 \\ 0 & d & K_B & J_W & 0 \\ b & 0 & e & K_U & J_R \\ c & 0 & 0 & f & K_S \end{bmatrix}$$

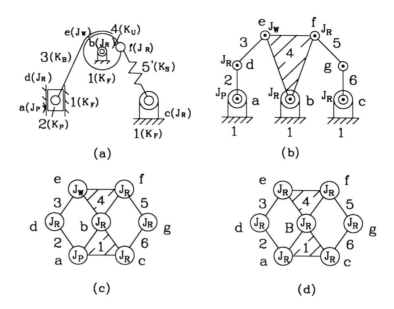

Figure 7.9 Generalization of a slider-pulley-spring mechanism

The ground link (K_F) is generalized into ternary link 1, the slider (K_P) is generalized into binary link 2, the belt (K_B) is generalized into binary link 3, the pulley (K_U) is generalized into ternary link 4, and the spring (K_S) is generalized into a dyad (binary links 5 and 6). Prismatic pair a is generalized into a revolute joint, and wrapping joint e is also generalized into a revolute joint. The corresponding generalized mechanical device is shown in Figure 7.9(b). Since there is no multiple joint in the device, the corresponding generalized chain is obtained by releasing the ground link as shown in Figure 7.9(c). And Figure 7.9(d) is its corresponding generalized kinematic chain. This is a Watt-chain with six links and seven revolute joints.

[Example 7.2]
An ordinary gear train, Figure 7.10(a).
This is a (4,5) mechanism. The four members are: the ground link (member 1, K_F), gear 1 (member 2, K_{G1}), gear 2 (member 3, K_{G2}), and gear 3 (member 4, K_{G3}). And the five joints are: joint a (members 1 and 2; J_R), joint b (members 1 and 3; J_R), joint c (members 1 and 4; J_R), joint d (members 2 and 3; J_G), and joint e (members 3 and 4; J_G). Its topology matrix, M_T, is:

$$M_T = \begin{bmatrix} K_F & J_R & J_R & J_R \\ a & K_{G1} & J_G & 0 \\ b & d & K_{G2} & J_G \\ c & 0 & e & K_{G3} \end{bmatrix}$$

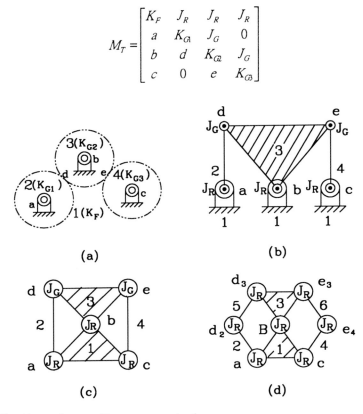

Figure 7.10 **Generalization of an ordinary gear train**

The ground link (K_F) is generalized into ternary link 1, gear 1 (K_{G1}) is generalized into binary link 2, gear 2 (K_{G2}) is generalized into ternary link 3, and gear 3 (K_{G3}) is generalized into binary link 4. The corresponding generalized mechanical device is shown in Figure 7.10(b), and the corresponding generalized chain is shown in Figure 7.10(c). This generalized chain can further be transformed into a generalized kinematic chain, as shown in Figure 7.10(d), by replacing generalized gear joints d and e by binary links 5 and 6 with generalized revolute joints d_2 and d_3, and e_3 and e_4 at both ends, respectively. It is a (4,5) generalized chain. It is also a Watt-chain with six links and seven revolute joints.

[Example 7.3]
A cam-roller-actuator mechanism, Figure 7.11(a).

This is a (4,5) mechanism. The four members are: the ground link (member 1, K_F), the cam (member 2, K_A), the roller (member 3, K_O), and the actuator (member 4', K_T). And the five joints are: joint a (members 1 and 2; J_R), joint b (members 1 and 3; J_O), joint c (members 1 and 4'; J_R), joint d (members 2 and 3; J_A), and joint e (members 3 and 4'; J_R). Its topology matrix, M_T, is:

$$M_T = \begin{bmatrix} K_F & J_R & J_O & J_R \\ a & K_A & J_A & 0 \\ b & d & K_O & J_R \\ c & 0 & e & K_T \end{bmatrix}$$

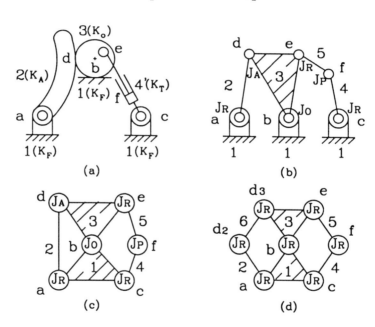

Figure 7.11 Generalization of a cam-roller-actuator mechanism

The ground link (K_F) is generalized into ternary link 1, the cam (K_A) is generalized into binary link 2, the roller (K_O) is generalized into ternary link 3, and the actuator (K_T) is generalized into the piston (binary link 4) and the cylinder (binary link 5) incident to prismatic pair f. The corresponding generalized mechanical device and chain are shown in Figures 7.11(b) and (c), respectively. This generalized chain can further be generalized into a generalized kinematic chain, as shown in Figure 7.11(d), by replacing generalized rolling joint b by generalized revolute joint b, generalized cam joint d by binary link 6 with generalized revolute joints d_2 and d_3 at both ends, and generalized prismatic joint f by generalized revolute joint f. It is a

(5,6) generalized chain. Furthermore, it is a (6,7) generalized kinematic chain, and it is also a Watt-chain with six links and seven revolute joints.

[Example 7.4]
A spring-loaded clamping device, Figure 7.12(a).

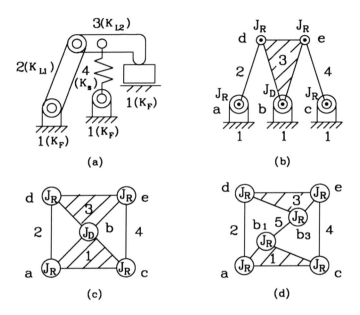

Figure 7.12 Generalization of a spring-loaded clamping device

This is a (4,5) mechanism. The four members are: the ground link (member 1, K_F), the pivot link (member 2, K_{L1}), the clamping link (member 3, K_{L2}), and the spring (member 4, K_S). And the five joints are: joint a (members 1 and 2; J_R), joint b (members 1 and 3; J_D), joint c (members 1 and 4; J_R), joint d (members 2 and 3; J_R), and joint e (members 3 and 4; J_R). Its topology matrix, M_T, is:

$$M_T = \begin{bmatrix} K_F & J_R & J_D & J_R \\ a & K_{L1} & J_R & 0 \\ b & d & K_{L2} & J_R \\ c & 0 & e & K_S \end{bmatrix}$$

The ground link (K_F) is generalized into ternary link 1, the pivot link (K_{L1}) is generalized into binary link 2, the clamping link (K_{L2}) is generalized into ternary link 3, and the spring (K_S) is generalized into binary link 4. The corresponding generalized device and chain are shown in Figures 7.12(b) and (c), respectively. This generalized chain can further be transformed into a generalized kinematic chain, as shown in Figure 7.12(d), by replacing generalized direct contact b by binary link 5 with generalized revolute joints b_1 and b_3 at both ends. It is a (4,5) generalized chain or a (5,6) generalized kinematic chain.

[Example 7.5]
A Swash-plate mechanism, Figure 7.13(a).

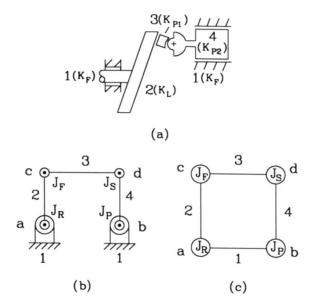

Figure 7.13 **Generalization of a Swash-plate mechanism**

This is a (4,4) mechanism. The four members are: the ground link (member 1, K_F), the plate (member 2, K_L), slider 1 (member 3, K_{P1}), and slider 2 (member 4, K_{P2}). And the four joints are: joint a (members 1 and 2; J_R), joint b (members 1 and 4; J_P), joint c (members 2 and 3; J_F), and joint d (members 3 and 4; J_S). Its topology matrix, M_T, is:

$$M_T = \begin{bmatrix} K_F & J_R & 0 & J_P \\ a & K_L & J_F & 0 \\ 0 & c & K_{P1} & J_S \\ b & 0 & d & K_{P2} \end{bmatrix}$$

The ground link (K_F) is generalized into binary link 1, the plate (K_L) is generalized into binary link 2, slider 1 (K_{P1}) is generalized into binary link 3, and slider 2 (K_{P2}) is generalized into binary link 4. The corresponding generalized mechanical device and generalized chain are shown in Figures 7.13(b) and (c), respectively. It is a (4, 4) generalized chain.

[Example 7.6]
A front wheel suspension mechanism, Figure 7.14(a).

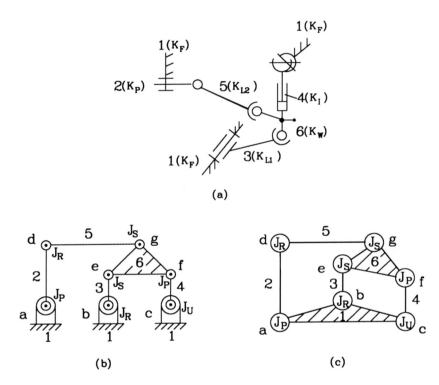

Figure 7.14 Generalization of a front wheel suspension mechanism

This is a (6,7) mechanism. The six members are: the ground link (member 1, K_F), the steering slider (member 2, K_P), the pivot link (member 3, K_{L1}), the piston (member 4, K_I), the connecting link (member 5, K_{L2}), and the cylinder that is also the wheel link (member 6, K_W). And the seven joints are: joint a (members 1 and 2; J_P), joint b (members 1 and 3; J_R), joint c (members 1 and 4; J_U), joint d (members 2 and 5; J_R), joint e (members 3 and 6; J_S), joint f (members 4 and 6; J_P), and joint g (members 5 and 6; J_S). Its topology matrix, M_T, is:

$$M_T = \begin{bmatrix} K_F & J_P & J_R & J_U & 0 & 0 \\ a & K_P & 0 & 0 & J_R & 0 \\ b & 0 & K_{L1} & 0 & 0 & J_S \\ c & 0 & 0 & K_I & 0 & J_P \\ 0 & d & 0 & 0 & K_{L2} & J_S \\ 0 & 0 & e & f & g & K_W \end{bmatrix}$$

The ground link (K_F) is generalized into ternary link 1, the steering slider (K_P) is generalized into binary link 2, the pivot link (K_{L1}) is generalized into binary link 3, the piston (K_I) is generalized into binary link 4, the connecting link (K_{L2}) is generalized into binary link 5, and the wheel link (K_W) is generalized into ternary link 6. The corresponding generalized mechanical device and generalized chain are shown in Figures 7.14(b) and (c), respectively. It is a (6,7) generalized chain.

7.6 Summary

Generalization is one of the major steps of the creative design methodology presented in Chapter 6. Mechanical devices involving various types of elements are transformed into corresponding generalized (kinematic) chains with only generalized (revolute) joints and generalized (kinematic) links. During the process of transforming a mechanical device into its corresponding generalized (kinematic) chain, the topological incidency and adjacency among members and joints and the number of degrees of freedom should be the same.

Through the process of generalization, design engineers are able to study and compare various devices in a very basic way. And mechanical devices with different topological structures may have identical generalized forms.

Problems

7.1 For the slider-crank mechanism shown in Figure 4.13(b), identify its corresponding generalized kinematic chain.

7.2 For the scooter anti-dive suspension mechanism shown in Figure 4.12, identify its corresponding generalized kinematic chain.

7.3 For the aircraft horizontal tail control mechanism shown in Figure 5.5, identify its corresponding generalized kinematic chain.

7.4 For the automotive front suspension mechanism shown in Figure 2.8, identify its corresponding generalized chain and generalized kinematic chain.

7.5 For the push-pull tool mechanism shown in Figure 3.2, identify its corresponding generalized chain and generalized kinematic chain.

7.6 For the hybrid motorcycle transmission shown in Figure 5.1, identify its corresponding generalized chain and generalized kinematic chain.

7.7 For the transmission mechanism of the rapier type shuttleless loom shown in Figure 4.2, identify its corresponding generalized chain and generalized kinematic chain.

7.8 Identify three different planar devices that have the same generalized kinematic chain.

7.9 Identify two different spatial devices that have the same generalized chain or generalized kinematic chain.

References

Franke, R., Vom Aufbau der Getribe, VDI-Verlag, Dusseldorf, 1958.

Hall, A. S. Jr., Generalized Linkages Forms of Mechanical Devices, ME261 class notes, Purdue University, West Lafayette, Indiana, spring 1978.

Johnson, R. C. and Towfigh, K., "Creative design of epicyclic gear trains using number synthesis," ASME Transactions, *Journal of Engineering for Industry*, May 1967, pp. 309-314.

Johnson, R. C., "Design synthesis aids to creative thinking," *Machine Design*, November 1973, pp. 158-163.

Miller, S., "Structural analysis of kinematic configurations with rigid, yielding, liquid and gas members," *Mechanism and Machine Theory*, Vol. 20, No. 3, 1985, pp. 209-213.

Soni, A. H., Mechanism Synthesis and Analysis, McGraw-Hill, 1974.

Yan, H. S., "A methodology for creative mechanism design," *Mechanism and Machine Theory*, Vol. 27, No. 3, 1992, pp. 235-242.

Yan, H. S. and Hwang, Y. W., "The generalization of mechanical devices," *Journal of the Chinese Society of Mechanical Engineers (Taiwan)*, Vol.9, No. 4, 1988, pp. 283-293.

CHAPTER 8

GENERALIZED CHAINS

Once a mechanical device is transformed into its corresponding generalized chain, the next step of the creative design methodology is to synthesize all possible chains with the required numbers of links and joints. The purposes of this and the following chapters are to provide various atlases of generalized chains and kinematic chains, respectively, as the inclusive data banks for the generation of all possible design concepts.

8.1 Generalized Chains

A *generalized chain* consists of generalized links connecting by generalized joints. It is connected, closed, without any bridge-link, and with simple joints only. An (N_L, N_J) generalized chain refers to a generalized chain with N_L generalized links and N_J generalized joints. The topological structure of a generalized chain is characterized by the number and the type of links, the number of joints, and the incidences between links and joints, and can be represented by its topology matrix, M_T, as defined in Chapter 2.

Each joint in a generalized chain is a generalized joint, i.e., the type of joint is not specified. If all joints in a generalized chain are specified, the number of degrees of freedom is positive, and the motion of this chain with one member grounded is constrained, it becomes a *kinematic chain*. It is a *rigid chain* if the number of degrees of freedom is not positive. Figure 8.1 shows the relationships among generalized chains, kinematic chains, and rigid chains.

8. Generalized Chains

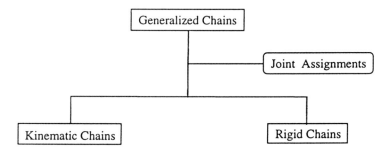

Figure 8.1 Generalized, kinematic, and rigid chains

For the (3,3) generalized chain shown in Figure 8.2(a), if joints a and b are revolute pairs and joint c is a cam pair, as shown in Figure 8.2(b), based on Equation (2.1), $N_L=3$, $C_{pR}=2$, $N_{JR}=2$, $C_{pA}=1$, $N_{JA}=1$, the degrees of freedom F_p of this planar device is:

$$F_p = 3(N_L-1) - (N_{JR}C_{pR}+N_{JA}C_{pA})$$
$$= (3)(3-1) - [(2)(2)+(1)(1)]$$
$$= 1$$

It is a (3,3) kinematic chain with one degree of freedom. If all three joints are revolute joints, as shown in Figure 8.2(c), based on Equation (2.1), $N_L=3$, $C_{pR}=2$, $N_{JR}=3$, the F_p is:

$$F_p = 3(N_L-1) - (N_{JR}C_{pR})$$
$$= (3)(3-1) - [(3)(2)]$$
$$= 0$$

It is a (3,3) rigid chain with zero degrees of freedom.

Figure 8.3(a) shows a generalized chain with five links (links 1, 2, 3, 4, and 5) and six joints (joints a, b, c, d, e, and f). If joints a, b, d, e, and f are revolute pairs, joint c is a cam pair, and link 1 is grounded, based on Equation (2.1), $N_L=5$, $C_{pR}=2$, $N_{JR}=5$, $C_{pA}=1$, $N_{JA}=1$, the F_p is:

$$F_p = 3(N_L-1) - (N_{JR}C_{pR}+N_{JA}C_{pA})$$
$$= (3)(5-1) - [(5)(2)+(1)(1)]$$
$$= 1$$

It is a planar five-bar mechanism with one degree of freedom, as shown in Figure 8.3(b). If joints a, b, c, and e are revolute pairs, joints d and f are gear pairs, and link 1 is grounded, based on Equation (2.1), $N_L=5$, $C_{pR}=2$, $N_{JR}=4$, $C_{pA}=1$, $N_{JA}=2$, the F_p is:

8.1. Generalized Chains • 119

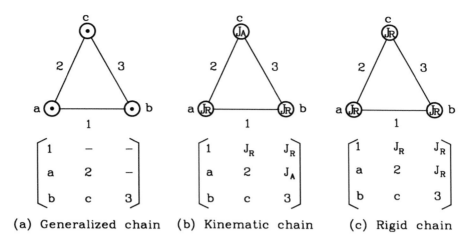

(a) Generalized chain (b) Kinematic chain (c) Rigid chain

Figure 8.2 Types of the (3,3) chain

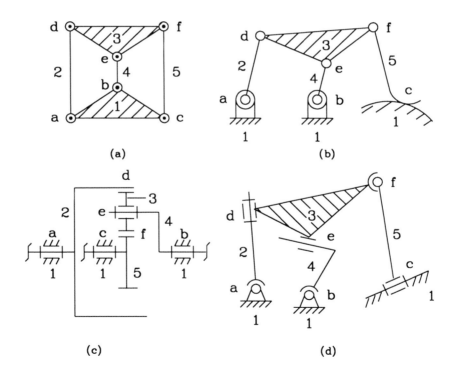

Figure 8.3 A (5,6) generalized chain and its derived mechanisms

$$F_p = 3(N_L\text{-}1) - (N_{JR}C_{pR}+N_{JA}C_{pA})$$
$$= (3)(5\text{-}1) - [(4)(2)+(2)(1)]$$
$$= 2$$

It is a five-bar planetary gear train with two degrees of freedom, as shown in Figure 8.3(c). If joints a, b, and f are spherical pairs, joints c and d are revolute pairs, joint e is a cylindrical pair, and link 1 is grounded, based on Equation (2.2), $N_L=5$, $C_{sR}=5$, $N_{JR}=2$, $C_{sC}=4$, $N_{JC}=1$, $C_{sS}=3$, $N_{JS}=3$, the degrees of freedom F_s for this spatial device is:

$$F_s = 6(N_L\text{-}1) - (N_{JR}C_{sR}+N_{JC}C_{sC}+N_{JS}C_{sS})$$
$$= (6)(5\text{-}1) - [(2)(5)+(1)(4)+(3)(3)]$$
$$= 1$$

It is a spatial five-bar mechanism with one degree of freedom.

Therefore, the atlases of kinematic chains and rigid chains can be identified from the atlas of generalized chains.

8.2 Link Assortments

The *link assortment*, A_L, of a generalized chain is the number and the type of links in the chain. It is a set of numbers consisting of the numbers of binary links (N_{L2}), ternary links (N_{L3}), quaternary links (N_{L4}), ... etc., and is expressed as:

$$A_L = [N_{L2}/N_{L3}/N_{L4}/...]$$

Since a generalized chain must be connected, closed, and without any bridge-link, the link assortments of a generalized chain with N_L links and N_J joints can be obtained by solving the following two equations:

$$N_{L2} + N_{L3} + ... + N_{Li} + ... + N_{Lm} = N_L \qquad (8.1)$$
$$2N_{L2} + 3N_{L3} + ... + iN_{Li} + ... + mN_{Lm} = 2N_J \qquad (8.2)$$

where N_{Li} is the number of links with i incident joints and m is the maximum number of joints incident to a link. Furthermore, the number of joints N_J is constrained by the following expression:

$$N_L \leq N_J \leq N_L(N_L\text{-}1)/2 \qquad (8.3)$$

The maximum value of m, i.e., m_{max}, can be derived based on elementary concepts of graph theory, and is expressed as:

$$m_{max} = \begin{cases} N_J - N_L + 2 & \text{for } N_L \leq N_J \leq 2N_L-3 \\ N_L - 1 & \text{for } 2N_L-3 \leq N_J \leq N_L(N_L-1)/2 \end{cases} \quad (8.4)$$

Based on Equations (8.1)-(8.4), all possible link assortments of generalized chains can be obtained.

[Example 8.1]
List link assortments of (6,7) generalized chains.

For (6,7) generalized chains, $N_L=6$, $N_J=7$, based on Equation (8.4), m_{max} is:

$$m_{max} = N_J - N_L + 2$$
$$= 7-6+2$$
$$= 3$$

Therefore, Equations (8.1) and (8.2) become:

$$N_{L2} + N_{L3} = 6$$
$$2N_{L2} + 3N_{L3} = 14$$

By solving these two equations, $N_{L2}=4$ and $N_{L3}=2$, i.e., the corresponding link assortment is:

$$A_L=[4/2]$$

as shown in Figure 8.4(a).

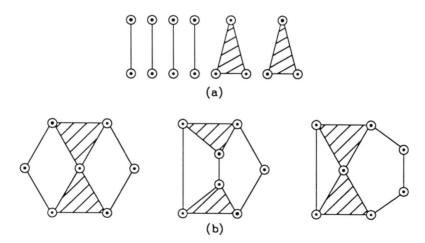

Figure 8.4 Atlas of (6,7) generalized chains with $A_L=[4/2]$

[Example 8.2]

List link assortments of generalized chains with four links.

For generalized chains with four links ($N_L=4$), based on Equation (8.3), the numbers of joints (N_J) can be 4, 5, and 6. And based on Equation (8.4), for $N_J=4$, m_{max} is:

$$m_{max} = N_J - N_L + 2$$
$$= 4-4+2$$
$$= 2$$

for $N_J=5$, m_{max} is:

$$m_{max} = N_J - N_L + 2$$
$$= 5-4+2$$
$$= 3$$

and, for $N_J=6$, m_{max} is:

$$m_{max} = N_L - 1$$
$$= 4-1$$
$$= 3$$

Therefore, for $N_J=4$ and $m_{max}=2$, Equations (8.1) and (8.2) become:

$$N_{L2} = 4$$
$$2N_{L2} = 8$$

i.e., $N_{L2}=4$ and the corresponding link assortment is $A_L=[4]$, as shown in Figure 8.5(a).

For $N_J=5$ and $m_{max}=3$, Equations (8.1) and (8.2) become:

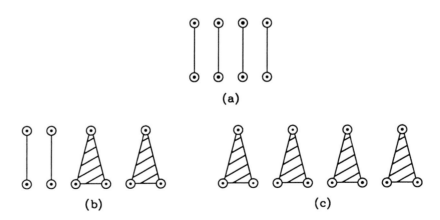

Figure 8.5 Link assortments of generalized chains with four links

$$N_{L2} + N_{L3} = 4$$
$$2N_{L2} + 3N_{L3} = 10$$

i.e., $N_{L2}=2$, $N_{L3}=2$, and the corresponding link assortment is $A_L=[2/2]$, as shown in Figure 8.5(b).

For $N_J=6$ and $m_{max}=3$, Equations (8.1) and (8.2) become:

$$N_{L2} + N_{L3} = 4$$
$$2N_{L2} + 3N_{L3} = 12$$

i.e., $N_{L2}=0$, $N_{L3}=4$, and the corresponding link assortment is $A_L=[0/4]$, as shown in Figure 8.5(c).

The atlas of (N_L, N_J) generalized chains can be obtained by assembling link assortments with N_L links and N_J joints. A particular link assortment might be assembled in more than one way, yielding different generalized chains. When assembling a link assortment to form its corresponding generalized chains, the following restrictions must be satisfied:
1. All links must be used to make the chain connected.
2. All joints must be used to make the chain closed.
3. No bridge-link should be formed.
4. A joint can only have two incident links to make the chain with simple joints only.
5. No link shall be attached to another by more than one joint.

For the link assortment of $A_L=[4/2]$ shown in Figure 8.4(a), three (6,7) generalized chains can be assembled as shown in Figure 8.4(b). The first one is called Watt-chain, and the second one is call Stephenson-chain.

8.3 Graphs and Chains

Ever since the early 1960s, the concepts of graph theory have been studied and applied for the structural analysis and synthesis of various types of chains and mechanisms. In order to utilize graph theory as a mathematical model for the structure of chains, some definitions on graphs are needed to be described first.

A *graph* $G=(S_N, S_E)$ may be defined as a set S_N of p nodes where S_N is finite and not empty, with another set S_E of q edges, each edge being a set of two distinct nodes. A graph is said to be *connected* if every pair of nodes is joined by a path. A graph is *closed* if it is connected and every node has at least two incident edges. A graph is a *block* if it is connected and has no bridge-nodes. A *bridge-node* in a graph is a node whose removal results in a disconnected graph. A *planar block* is a block that can be drawn in the plane with no crossings. A graph with p nodes and q edges is called a (p,q) graph. Graphically, a node is symbolized by a dark dot and an edge by a line. Figure 8.6 shows all the eleven graphs having four nodes. The first five graphs (G_1-G_5) are disconnected, the last six graphs (G_6-G_{11}) are connected, and the last three graphs (G_9-G_{11}) are blocks. The lower left node of G_8 is a bridge-node as its removal result in a disconnected subgraph (with three nodes). And G_9, G_{10}, and G_{11} are planar blocks.

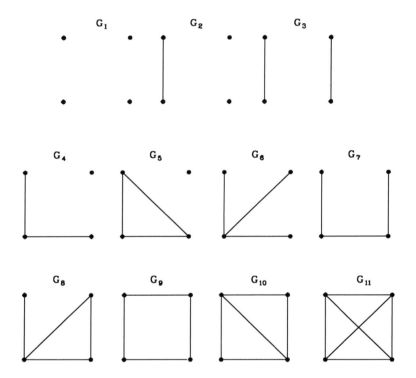

Figure 8.6 Atlas of graphs with four nodes

For a graph G with no isolated nodes, its *line graph* G_L has the edges of G as the node set. The two nodes p_1 and p_2 of G_L are adjacent when the following holds: regarding each edge q_i of G as the set consisting of the two nodes that it joins, the intersection $p_1 \cap p_2$ is a singleton, i.e., nodes p_1 and p_2 of G_L are joined by an edge in G_L if edges q_1 and q_2 of G are incident with just one common node of G. For the $(4,5)$ graph shown in Figure 8.7(a), its corresponding line graph is shown in Figure 8.7(b).

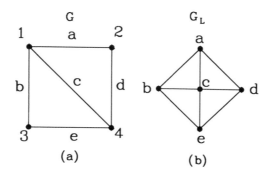

Figure 8.7 Line graph of the (4,5) graph

Based on some concepts from the theory of graphs and hypergraphs, generalized chains from planar blocks can be constructed according to the following algorithm:

Step 1. For each given (p,q) planar block G, draw G in the plane with no crossing.
Step 2. For each node, list those edges incident with the node.
Step 3. Construct line graph G_L.
Step 4. Replace each node of G_L by a small circle with a dot in the center.
Step 5. Replace each complete subgraph of G_L that is determined by a node of G with at least three incident edges by a hatched polygon. This is done by removing the interior edges (when the incident edge is four or more) to obtain the perimeter polygon, and then hatching the interior of this polygon.

The resulting configuration is the corresponding generalized chain of the graph.

Furthermore, there exists a one-to-one correspondence between generalized chains and (planar) blocks, i.e., every generalized chain has a unique associated graph that is a (planar) block and each (planar) block produces a unique generalized chain. This correspondence has the following consequences:

1. Every theorem about (planar) blocks transforms into a statement that is valid for all generalized chains.
2. Every concept or numerical invariant associated with (planar) blocks has a corresponding meaning for generalized chains, and conversely.
3. Every law that applies to all generalized chains can be converted into a theorem concerning (planar) blocks.

For the (6,7) planar block shown in Figure 8.8(a), its corresponding generalized chain can be constructed as shown in Figure 8.8(b).

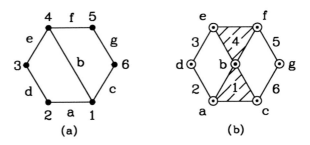

Figure 8.8 A (6,7) planar block and its corresponding generalized chain

8.4 Numbers of Generalized Chains

The one-to-one correspondence between (planar) blocks and generalized chains as discussed in the previous section enables the adoption of results in graph theory for applications in the number synthesis of generalized chains. Although the enumeration of planar graphs and of planar blocks remains as extremely difficult unsolved problems, the numbers of blocks with various nodes are available. Table 8.1 lists the total number of blocks (n_b) up to ten nodes, and Table 8.2 lists the numbers of (p,q) blocks (n_g) up to seven nodes. Table 8.2 also provides the numbers of (N_L, N_J) generalized chains whose corresponding blocks are not necessarily planar.

8.5 Atlas of Generalized Chains

The atlas of various (N_L, N_J) generalized chains can be synthesized by assembling the corresponding link assortments. It can also be obtained by transforming from available atlases of (planar) blocks.

Table 8.3, in which N_{GC} denotes the number of generalized chains, and Figures 8.9-8.23 show some important atlases of generalized chains that should cover most industrial applications.

8.6 Summary

A generalized chain consists of generalized links connecting by generalized joints. If all joints in a generalized chain are specified, the number of degrees of freedom is positive, and the motion of this chain with one member grounded is constrained, it becomes a kinematic chain. And it is a rigid chain if the number of degrees of freedom is not positive.

8.4. Numbers of Generalized Chains

Table 8.1 Numbers of blocks (n_b) up to ten nodes

N_L	n_b
3	1
4	3
5	10
6	56
7	468
8	7,123
9	194,066
10	9,743,542

Table 8.2 Numbers of (p,q) blocks (n_g) up to seven nodes

p	q	n_g	p	q	n_g	p	q	n_g
3	3	1	6	8	9	7	11	82
4	4	1	6	9	14	7	12	94
4	5	1	6	10	12	7	13	81
4	6	1	6	11	8	7	14	59
5	5	1	6	12	5	7	15	38
5	6	2	6	13	2	7	16	20
5	7	3	6	14	1	7	17	10
5	8	2	6	15	1	7	18	5
5	9	1	7	7	1	7	19	2
5	10	1	7	8	4	7	20	1
6	6	1	7	9	20	7	21	1
6	7	3	7	10	50			

Table 8.3 Atlas of generalized chains

N_L	N_J	N_{GC}	Figure
3	3	1	8.9
4	4-6	3	8.10
5	5-10	10	8.11
6	6	1	8.12
6	7	3	8.13
6	8	9	8.14
6	9	14	8.15
7	7	1	8.16
7	8	4	8.17
7	9	20	8.18
7	10	50	8.19
8	8	1	8.20
8	9	6	8.21
8	10	40	8.22
8	11	77	8.23

128 ▪ 8. Generalized Chains

Figure 8.9   Atlas of generalized chains with three links

Figure 8.10   Atlas of generalized chains with four links

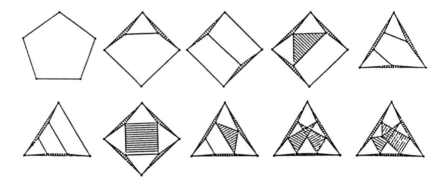

Figure 8.11   Atlas of generalized chains with five links

Figure 8.12   Atlas of (6,6) generalized chain

Figure 8.13   Atlas of (6,7) generalized chains

8.5. Atlas of Generalized Chains • 129

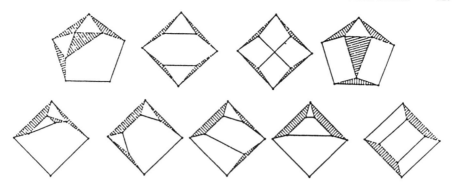

Figure 8.14 Atlas of (6,8) generalized chains

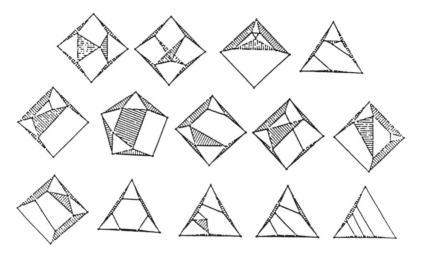

Figure 8.15 Atlas of (6,9) generalized chains

Figure 8.16 Atlas of (7,7) generalized chain

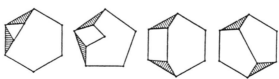

Figure 8.17 Atlas of (7,8) generalized chains

130 ■ 8. Generalized Chains

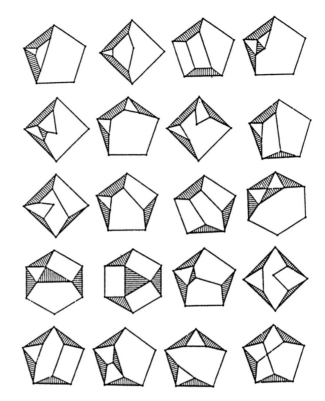

Figure 8.18 Atlas of (7,9) generalized chains

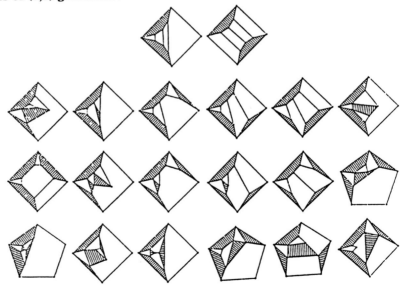

8.5. Atlas of Generalized Chains • 131

Figure 8.19 Atlas of (7,10) generalized chains

Figure 8.20 Atlas of (8,8) generalized chain

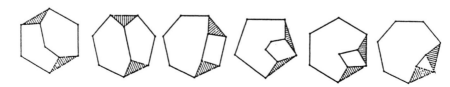

Figure 8.21 Atlas of (8,9) generalized chains

132 ▪ 8. Generalized Chains

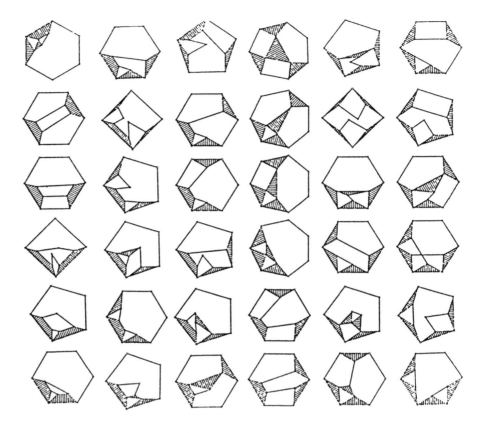

Figure 8.22 Atlas of (8,10) generalized chains

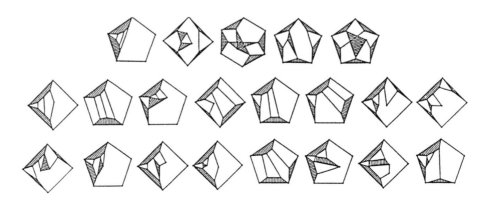

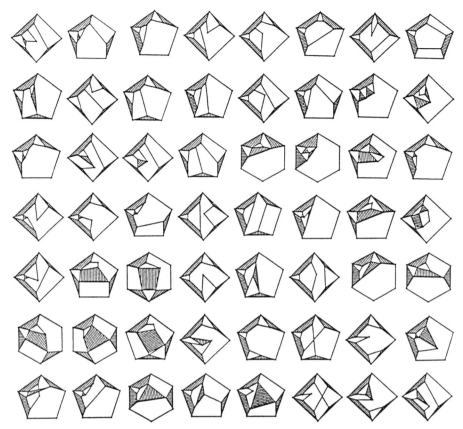

Figure 8.23 Atlas of (8,11) generalized chains

There exists a one-to-one correspondence between generalized chains and (planar) blocks. Every generalized chain has a unique associated graph that is a (planar) block, and each (planar) block produces a unique generalized chain.

The link assortments of a generalized chain is the number and the type of links in the chain. The atlas of various generalized chains can be synthesized by assembling the corresponding link assortments. It can also be obtained by transforming from available atlas of (planar) blocks.

The concept of generalized chains covers the application of kinematic chains for mechanisms and that of rigid chains for structures. The atlases of generalized chains listed in this chapter provide design engineers the necessary data bank for the generation of all possible topological structures of mechanical devices based on the creative design methodology presented in Chapter 6.

Problems

8.1 What are link assortments of generalized chains with eight links and ten joints?
8.2 What are link assortments of generalized chains with six links?
8.3 What are the corresponding generalized chains based on the assembly of link assortments obtained in Problem 8.2?
8.4 Sketch the corresponding chains for the atlas of graphs with four nodes shown in Figure 8.6.
8.5 Sketch the corresponding graphs for the atlas of generalized chains with six links and seven joints shown in Figure 8.4(b).
8.6 Sketch the corresponding graph for the aircraft horizontal tail control mechanism shown in Figure 2.7.
8.7 Identify one linkage mechanism and one cam mechanism that have the same generalized chain.
8.8 Identify one mechanism and one structure that have the same generalized chain.

References

Harary, F., Graph Theory, Addison-Wesley, 1969.
Harary, F. and Palmer, E. M., Graphical Enumeration, Academic Press, 1973.
Harary, F. and Yan, H. S., "Logical foundations of kinematic chains: graphs, line graphs, and hypergraphs," ASME Transactions, *Journal of Mechanical Design*, Vol. 112, No. 1, 1990, pp. 79-83.
Hwang, W. M., Computer-aided Structural Synthesis of Planar Kinematic Chains with Multiple Joints, Ph.D. dissertation, Department of Mechanical Engineering, National Cheng Kung University, Tainan, Taiwan, May 1984.
Hwang, W. M. and Yan, H. S., "Atlas of basic rigid chains," Proceedings of the 9th Applied Mechanisms Conference, Session IV-B, No. 1, Kansas City, Missouri, October 28-30, 1985.
Hwang, Y. W., An Expert System for Creative Mechanism Design, Ph.D. dissertation, Department of Mechanical Engineering, National Cheng Kung University, Tainan, Taiwan, May 1990.
Robinson, R. W., "Enumeration of nonseparable graphs," *Journal of Combinatorial Theory*, Vol. 9, 1970, pp. 327-356.
Yan, H. S., "A methodology for creative mechanism design," *Mechanism and Machine Theory*, Vol. 27, No. 3, 1992, pp. 235-242.
Yan, H. S. and Harary, F., "On the maximum value of the maximum degree of kinematic chains," ASME Transactions, *Journal of Mechanisms, Transmissions, and Automation in Design*, Vol. 109, No. 4, 1987, pp. 487-490.

CHAPTER 9

KINEMATIC CHAINS

In the process of creative mechanism design, the atlas of kinematic chains is needed. This chapter first describes how kinematic chains can be obtained from available generalized chains. Then, the definitions of various kinematic matrices and permutation groups are presented; and the algorithm for the enumeration of non-isomorphic kinematic chains based on these definitions is derived. Finally, some atlases of kinematic chains that are important for applications are provided.

9.1 Kinematic Chains

Atlases of kinematic chains can be identified from available atlases of generalized chains derived and provided in Chapter 8. If the type of each joint in a generalized chain is specified and the motion of this chain with one member grounded is constrained, it becomes a *kinematic chain*.

In general, a *simple kinematic chain* is referred to as a kinematic chain with revolute pairs and simple joints only. The atlas of kinematic chains with N_L links and N_J joints can be obtained from the atlas of (N_L, N_J) generalized chains by deleting those chains with three-bar loops or subchains with non-positive degrees of freedom. For example, for the atlas of three (6,7) generalized chains shown in Figure 9.1, the one shown in Figure 9.1(c) has a three-bar loop, the other two chains shown in Figures 9.1(a) and (b) are the atlas of (6,7) kinematic chains.

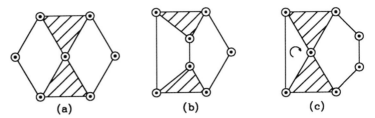

Figure 9.1 Atlas of (6,7) generalized chains

9.2 Rigid Chains

A *basic rigid chain* is a rigid chain without any rigid subchain. The simplest (basic) rigid chain is the (3,3) chain, as shown in Figure 9.2. The (5,6) chain shown in Figure 9.3(a), with zero degrees of freedom, is a five-bar rigid chain. Since it contains a three-bar basic rigid chain (links 1-2-3), it is not a five-bar basic rigid chain. However, the (5,6) chain shown in Figure 9.3(b) is a basic rigid chain.

Figure 9.2 Three-bar (basic) rigid chain

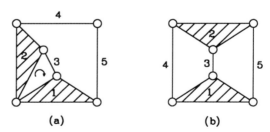

Figure 9.3 Two five-bar rigid chains

A kinematic chain with positive degrees of freedom and containing any basic rigid chain is a *degenerate kinematic chain*. A degenerate kinematic chain can be transformed into a chain with fewer links by replacing the basic rigid chain with a single link.

For example, the (10,13) kinematic chain shown in Figure 9.4(a) contains a seven-bar basic rigid chain as shown in Figure 9.4(b), and is degenerated into the (4,4) kinematic chain as shown in Figure 9.4(c).

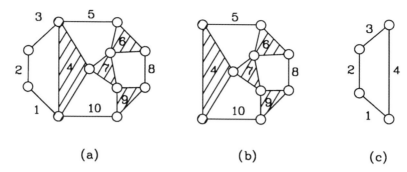

Figure 9.4 **Degeneration of a (10,13) kinematic chain**

Atlases of non-isomorphic kinematic chains can also be synthesized based on the concept of permutation groups. For this purpose, a number of definitions and terminology involving kinematic matrices and permutation groups are required.

9.3 Kinematic Matrices

The concept of matrices is a powerful tool for the representation of the topological structures of various chains. The definitions of link adjacency matrix, labeled link adjacency matrix, and contracted link adjacency matrix are now introduced.

Link adjacency matrix

The *link adjacency matrix*, M_{LA}, of a generalized (kinematic) chain with N_L links and N_J joints is an $N_L \times N_L$ matrix with its elements $e_{ij}=1$ if link i is adjacent to link j, and $e_{ij}=0$ otherwise.

For the (6,7) Watt-chain shown in Figure 9.5(a), its link adjacency matrix, M_{LA}, is:

$$M_{LA} = \begin{bmatrix} 0 & 1 & 0 & 1 & 0 & 1 \\ 1 & 0 & 1 & 0 & 0 & 0 \\ 0 & 1 & 0 & 1 & 0 & 0 \\ 1 & 0 & 1 & 0 & 1 & 0 \\ 0 & 0 & 0 & 1 & 0 & 1 \\ 1 & 0 & 0 & 0 & 1 & 0 \end{bmatrix}$$

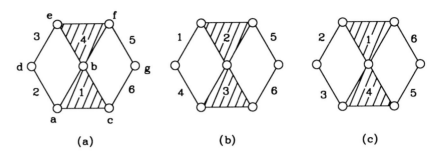

Figure 9.5 Labeled Watt-chains

Labeled link adjacency matrix

A *labeled chain* is a chain with its links being labeled by integers {1,2,3,...,i,...} and joints by letters {a,b,c,...,k,...}. The *labeled link adjacency matrix*, M_{LLA}, of a labeled generalized (kinematic) chain with N_L links and N_J joints is an $N_L \times N_L$ matrix with its elements $e_{ii}=i$ for the ith link, $e_{ij}=k$ if joint k is incident to links i and j, and $e_{ij}=0$ otherwise.

For the labeled (6,7) Watt-chain shown in Figure 9.5(a), its labeled link adjacency matrix, M_{LLA}, is:

$$M_{LLA} = \begin{bmatrix} 1 & a & 0 & b & 0 & c \\ a & 2 & d & 0 & 0 & 0 \\ 0 & d & 3 & e & 0 & 0 \\ b & 0 & e & 4 & f & 0 \\ 0 & 0 & 0 & f & 5 & g \\ c & 0 & 0 & 0 & g & 6 \end{bmatrix}$$

Contracted link adjacency matrix

A string of binary links in a kinematic chain is regarded as a *contract link*. For a *contracted link adjacency matrix*, M_{CLA}, its diagonal element e_{ii}, called *link element*, indicates the type of link i. The value of e_{ii} is defined such that $e_{ii}=+u$ if link i is a multiple link with u joints and $e_{ii}=-v$ if link i is a contracted link with v binary links. The off-diagonal element e_{ij}, called *joint element*, indicates the number of joints incident between link i and link j. The value of e_{ij} is defined such that $e_{ij}=w$ if link i and link j are connected by w joints. Note that w can only be 0, 1, or 2. The case of $w=2$ happens only when one of these two adjacent links is a contracted link with three or more binary links, the other one is a multiple link, and both ends of the contracted link are connected to this multiple link.

9.3. Kinematic Matrices

Without loss of generality, the diagonal elements of an M_{CLA} are arranged in a non-increasing sequence by exchanging the order of links. For example, the M_{CLA} matrix of the (10,13) kinematic chain shown in Figure 9.6(a) is expressed as:

$$M_{CLA} = \left[\begin{array}{cccc|cccc} 4 & 1 & 0 & 1 & 1 & 1 & 0 & 1 \\ 1 & 4 & 1 & 0 & 1 & 1 & 1 & 0 \\ 0 & 1 & 3 & 1 & 1 & 0 & 1 & 0 \\ 1 & 0 & 1 & 3 & 1 & 0 & 0 & 1 \\ \hline 1 & 1 & 0 & 0 & 1 & -2 & 0 & 0 \\ 0 & 1 & 1 & 0 & 1 & 0 & -2 & 0 \\ 1 & 0 & 0 & 1 & 1 & 0 & 0 & -2 \end{array}\right]$$

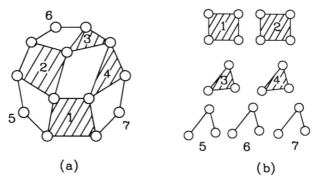

(a)   (b)

Figure 9.6 A (10,13) kinematic chain and its multiple and contracted links

in which diagonal elements imply that the kinematic chain has two quaternary links (links 1 and 2), two ternary links (links 3 and 4), and three contracted links with two binary links (links 5, 6, and 7), as shown in Figure 9.6(b).

For the sake of simplicity, an M_{CLA} matrix is divided into four submatrices, M_{ul}, M_{ur}, M_{ll}, and M_{lr}, as follows:

$$M_{CLA} = \left[\begin{array}{c|c} M_{ul} & M_{ur} \\ \hline M_{ll} & M_{lr} \end{array}\right]$$

Here, M_{ul}, the upper-left square submatrix in the M_{CLA}, expresses the configuration of multiple links. M_{lr} is the lower-right submatrix in the M_{CLA}, and the values of its off-diagonal elements are always zero because any two contracted links are non-adjacent. M_{ur}, the upper-right submatrix in the M_{CLA}, indicates the adjacent relation between multiple links and contracted links. And M_{ll}, the lower-left submatrix in the M_{CLA}, is the transport matrix of the M_{ur} matrix.

9.4 Permutation Groups

For a labeled generalized (kinematic) chain, three permutation groups, namely link-group, joint-group and chain-group, are defined according to some basic concepts from combinatorial theory.

A *permutation*, P, is a bisection (one-to-one and onto) of a finite set S into itself. The usual composition of mappings provides a binary operation for permutations on the same set. Furthermore, whenever a collection of permutations is closed with respect to this composition, it is called a *permutation group*, P_G. For example, the sequence (B,C,A,D) is a permutation of the set S=(A,B,C,D) in which A is transformed into B (A→B), B is transformed into C (B→C), C is transformed into A (C→A), and D is transformed into D (D→D). In this permutation, A→B→C→A forms a *cycle*, denoted by [ABC], with a length of three; and D→D forms another cycle [D] with a length of one. The cyclic representation of this permutation is denoted by P=[ABC][D].

Link-group

Let S_{LL}=(1,2,3,...) be the set of the labels of links of a generalized (kinematic) chain. Applying a permutation P of S_{LL} on the chain is equivalent to relabeling the links of the chain, and the original chain and the relabelled chain are isomorphic. For example, the permutation P=[13][24][5][6] transforms the chain shown in Figure 9.5(a) into the isomorphic one shown in Figure 9.5(b).

For some special permutations, the relabeled chain is the same as the original one, i.e., not only the link adjacency but also the labels of links are the same. In terms of graph theory, these two chains are *automorphic*. For example, the permutation P=[14][23][56] transforms the chain shown in Figure 9.5(a) into the automorphic one shown in Figure 9.5(c). The set of those special permutations that relabel the links of a chain and transform the chain into an automorphic one forms a group called *link-group* of the chain and denoted by D_L. For example, the link-group of the chain shown in Figure 9.5(a) is:

$$D_L = \{P_{L1}, P_{L2}, P_{L3}, P_{L4}\}$$

where

$$P_{L1} = [1][2][3][4][5][6]$$

$$P_{L2} = [1/4][2/3][5/6]$$
$$P_{L3} = [1][2/6][3/5][4]$$
$$P_{L4} = [1/4][2/5][3/6]$$

Joint-group

Let $S_{LJ}=(a,b,c,...)$ be the set of the labels of the joints of a generalized (kinematic) chain. Then, there exists a set of permutations that transforms the chain into an automorphic one. These permutations form a group called *joint-group*, D_J, of the chain. For example, the joint-group of the chain shown in Figure 9.5(a) is:

where
$$D_J = \{P_{J1}, P_{J2}, P_{J3}, P_{J4}\}$$

$$P_{J1} = [a][b][c][d][e][f][g]$$
$$P_{J2} = [a/e][b][c/f][d][g]$$
$$P_{J3} = [a/c][b][e/f][d/g]$$
$$P_{J4} = [a/f][b][c/e][d/g]$$

Chain-group

If the links and the joints of a generalized (kinematic) chain are labeled simultaneously, an analogous group called *chain-group*, D_C, is obtained. For example, the chain-group of the chain shown in Figure 9.5(a) is:

where
$$D_C = \{P_{C1}, P_{C2}, P_{C3}, P_{C4}\}$$

$$P_{C1} = [1][2][3][4][5][6][a][b][c][d][e][f][g]$$
$$P_{C2} = [1/4][2/3][5/6][a/e][b][c/f][d][g]$$
$$P_{C3} = [1][2/6][3/5][4][a/c][b][e/f][d/g]$$
$$P_{C4} = [1/4][2/5][3/6][a/f][b][c/e][d/g]$$

Similar class

Let $S=(s_1,s_2,s_3,...,s_k,...)$ be the set of the labels of the links (or joints) of a generalized (kinematic) chain. If there exists a permutation P of D_L (or D_J) that transforms s_i into s_j, the two elements s_i and s_j are *similar* subject to permutation P. Furthermore, the set S can be partitioned into some *similar classes* by putting similar elements into the same class. For the Watt-chain shown in Figure 9.5(a), {1,4} and {2,3,5,6} are two similar classes of links, and {b}, {a,c,e,f} and {d,g} are three similar classes of joints.

Permutation groups

The link-group of a generalized chain is obtained from its link adjacency matrix. The link adjacency matrices of isomorphic chains are equivalent if they differ by a link permutation of rows and by the same permutation of columns. However, the link adjacency matrices of automorphic chains are the same.

The algorithm for obtaining the link-group of a chain is as follows:
1. Characterize each link of the chain of interest by its attribute, i.e., the number of its incident joints and those of its adjacency links.
2. List all possible link permutations in which only those links with the same attribute can be relabeled with each other.
3. Apply each of the possible link permutations on the link adjacency matrix of the chain. If the resulting link adjacency matrix is the same as the original one, the permutation is the number of the link-group of the chain.

For the labeled (6,7) Stephenson-chain shown in Figure 9.7, both links 1 and 2 have three incident joints and all of their adjacency links have two incident joints. Therefore, links 1 and 2 have the same attribute, and so do links 3 and 4, and links 5 and 6. As a result, the number of possible link permutations is $2! \times 2! \times 2! = 8$, and they are:

$$P_{L1} = [1][2][3][4][5][6]$$
$$P_{L2} = [1/2][3][4][5][6]$$
$$P_{L3} = [1][2][3/4][5][6]$$
$$P_{L4} = [1][2][3][4][5/6]$$
$$P_{L5} = [1/2][3/4][5][6]$$
$$P_{L6} = [1/2][3][4][5/6]$$
$$P_{L7} = [1][2][3/4][5/6]$$
$$P_{L8} = [1/2][3/4][5/6]$$

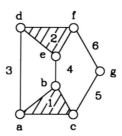

Figure 9.7 A labeled Stephenson-chain

Among these permutations, each one of P_{L1}, P_{L3}, P_{L6}, and P_{L8} transforms the link adjacency matrix into itself. Therefore, the link-group D_L of this chain is $\{P_{L1}, P_{L3}, P_{L6}, P_{L8}\}$.

By applying each link permutation of the labeled link adjacency matrix of the chain, the joint-group and chain-group are obtained by observing the transformations of non-diagonal elements of the labeled link adjacency matrix. For the Stephenson-chain shown in Figure 9.7, its corresponding labeled link adjacency matrix is:

$$\begin{bmatrix} 1 & 0 & a & b & c & 0 \\ 0 & 2 & d & e & 0 & f \\ a & d & 3 & 0 & 0 & 0 \\ b & e & 0 & 4 & 0 & 0 \\ c & 0 & 0 & 0 & 5 & g \\ 0 & f & 0 & 0 & g & 6 \end{bmatrix}$$

and the resulting labeled link adjacency matrix transformed by P_{L3}=[1][2][34][5][6] is:

$$\begin{bmatrix} 1 & 0 & b & a & c & 0 \\ 0 & 2 & e & d & 0 & f \\ b & e & 4 & 0 & 0 & 0 \\ a & d & 0 & 3 & 0 & 0 \\ c & 0 & 0 & 0 & 5 & g \\ 0 & f & 0 & 0 & g & 6 \end{bmatrix}$$

From the non-diagonal elements of these two labeled link adjacency matrices, it is found that a→b, b→a, c→c, d→e, e→d, f→f, and g→g. Therefore, there exist a joint permutation:

$$P_{J3} = [a/b][c][d/e][f][g]$$

and a chain permutation:

$$P_{C3} = [1][2][3/4][5][6][a/b][c][d/e][f][g]$$

corresponding to the link permutation P_{L3}.

9.5 Enumerating Algorithm

The atlas of non-isomorphic kinematic chains with only revolute pairs and simple joints can be synthesized according to the following algorithm, based on the concept of permutation groups. Since a contracted link adjacency matrix (M_{CLA}) represents the topological structure of a kinematic chain, the enumeration of kinematic chains can be achieved by constructing all possible contracted link adjacency matrices. The step by step algorithm for constructing all non-isomorphic contracted link adjacency matrices is described as followers (Figure 9.8):

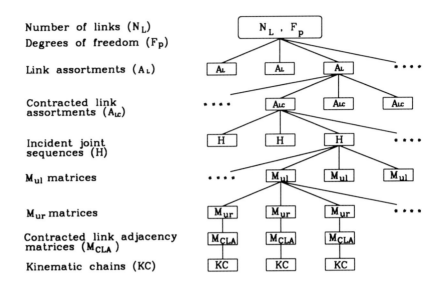

Figure 9.8   Enumerating algorithm of kinematic chains

Step 1. Input the numbers of links N_L and degrees of freedom F_p.

Step 2. Find link assortments $A_L=[N_{L2}/N_{L3}/.../N_{Li}/...]$ in which N_{Li} is the number of links with i incident joints.

Step 3. For each link assortment, find contracted link assortments $A_{LC}=[N_{c1}/N_{c2}/.../N_{ci}/...]$ in which N_{ci} is the number of contracted links with i binary links.

Step 4. For each contracted link assortment, find incident joint sequences $H=(a_1, a_2,...,a_i,...,a_{Nm})$ of multiple links, where N_m is the number of multiple links, and a_i is the number of joints incident between multiple link i and the other multiple links.

Step 5. For each incident joint sequence, find all non-isomorphic M_{ul} matrices as follows:
   (1) Find the permutation groups of joint elements of M_{ul} matrix.
   (2) Based on this permutation group, assign "0"s and "1"s to joint elements to construct all non-isomorphic M_{ul} matrices.

Step 6. For each generated M_{ul} matrix, find all non-isomorphic M_{ur} matrices as follows:
   (1) Find the permutation group of joint elements of M_{ur} matrix.
   (2) Based on this permutation group, assign "0"s, "1"s, and "2"s to joint elements to construct all non-isomorphic M_{ur} matrices.

Step 7. Construct corresponding M_{CLA} from generated M_{ur} matrices. Those M_{CLA} with rigid subchains are deleted.

Step 8. Transform each constructed M_{CLA} into its corresponding kinematic chain in graphic form.

In the following sections, each step is described in detail.

9.5.1 Step 1
Input the Numbers of Links and Degrees of Freedom

Since planar kinematic chains with revolute pairs and simple joints are our concern, the number of joints (N_J) can be calculated based on Equation (2.1), as follows:

$$N_J = [3(N_L-1) - F_p] / 2 \qquad (9.1)$$

in which F_p is the number of degrees of freedom and N_L is the number of links.

For a kinematic chain with ten links (N_L=10) and one degree of freedom (F_p=1), based on Equation (9.1), its number of joints N_J is 13, Figure 9.6(a).

9.5.2 Step 2
Find Link Assortments

Link assortment, $A_L=[N_{L2}/N_{L3}/N_{L4}/...]$, of a kinematic chain with N_L links and N_J joints can be obtained by solving the following three equations:

$$N_{L2} + N_{L3} + N_{L4} + ... + N_{Lm} = N_L \qquad (9.2)$$
$$2N_{L2} + 3N_{L3} + 4N_{L4} + ... + mN_{Lm} = 2N_J \qquad (9.3)$$

$$m = \begin{cases} (N_L-F_p+1)/2 & \text{if } F_p = 0 \text{ or } 1 \\ \min.\{(N_L-F_p+1), (N_L-F_p+1)/2\} & \text{if } F_p \geq 2 \end{cases} \qquad (9.4)$$

For kinematic chains with N_L=10 and F_p=1, and therefore N_J=13, all possible link assortments are: [4/6/0/0], [5/4/1/0], [6/2/2/0], [7/0/3/0], [6/3/0/1], [7/1/1/1], and [8/0/0/2]. The link assortment of the kinematic chain shown in Figure 9.6(a) is [6/2/2/0].

9.5.3 Step 3
Find Contracted Link Assortments

This step is equivalent to partition N_{L2} binary links into N_c parts, where N_c is the number of contracted links. Therefore, a contracted link assortment $A_{LC}= [N_{c1}/N_{c2}/.../N_{ci}/...]$ must satisfy the following equations:

$$N_{c1} + N_{c2} + \ldots + N_{cr} = N_c \qquad (9.5)$$
$$N_{c1} + 2N_{c2} + \ldots + rN_{cr} = N_{L2} \qquad (9.6)$$

It is obvious that a non-degenerate kinematic chain with F_p degrees of freedom cannot consist of a contracted link with F_p+2 or more binary links, therefore $r=F_p+1$. The range of N_c is:

$$J_m - J_m' \leq N_c \leq \min.\{N_{L2}, J_m\} \qquad (9.7)$$

where

$$2J_m = 3N_{L3} + 4N_{L4} + \ldots + qN_q \qquad (9.8)$$

$$2J_m' = \begin{cases} 0 & \text{if } N_m = 1 \\ 3(N_m-1) - 1 & \text{if } N_m = 2, 4, 6, \ldots \\ 3(N_m-1) - 2 & \text{if } N_m = 3, 5, 7, \ldots \end{cases} \qquad (9.9)$$

Based on Equations (9.5)-(9.9), all possible contracted link assortments can be generated for a given link assortment. For example, if $N_L=10$, $F_p=1$, and $A_L=[6/2/2/0]$, then $r=2$, $N_m=4$, $J_m=7$, $3<N_c<6$, and all possible contracted link assortments are: [0/3], [2/2], [4/1], and [6/0].

Once contracted link assortments are obtained, the value of e_{ii} of the M_{CLA} can be determined in a non-increasing sequence. For example, if $A_L=[6/2/2/0]$ and $A_{LC}=[2/2]$, then the two quaternary links must be labeled as links 1 and 2, the two ternary links as links 3 and 4, the two contracted links with one binary link as links 5 and 6, and the two contracted links with two binary links as 7 and 8. The sequence of the value of e_{ii} is (4,4,3,3,-1,-1,-2,-2).

9.5.4 Step 4
Find Incident Joint Sequences

Before constructing the M_{ul} matrix, all possible *incident joint sequences* $H=(a_1,a_2,\ldots,a_i,\ldots,a_{Nm})$ of multiple links must be found, where a_i is the number of joints incident between multiple link i and the other multiple links. In other words, a_i is the sum of the values of off-diagonal elements of row i in M_{ul} matrix, i.e.,

$$\sum_{\substack{j=1 \\ j \neq 1}}^{N_m} e_{ij} = a_i, \qquad i = 1, \ldots, N_m \qquad (9.10)$$

Equation (9.10) is called *compatibility constraints* of M_{ul} matrix. Furthermore, the sum of a_i, denoted by $2J_d$, can be obtained by the following expression:

$$2J_d = 2J_m - 2N_c \qquad (9.11)$$

The problem of determining incident joint sequences is equivalent to the problem of partitioning integer $2J_d$ into k parts and assigning these k parts into N_m elements of H sequence. Here, $k<N_m$, since some multiple links may not be adjacent to the other multiple links.

In order to avoid the configuration of multiple links with rigid subchains, the k parts must satisfy the following constraints:
1. The largest of parts, t, can not be greater than the maximum number of joints of multiple links, i.e., $t<e_{11}$, and must be less than the number of parts, i.e., $t<k$.
2. The degrees of freedom of the configuration of multiple links must be positive, i.e., $3(k-1)-2J_d>0$.

For example, if the sequence of link elements is (4,4,3,3,-1,-1,-2,-2), then $J_m=7$, $N_m=4$, $N_c=4$, $J_d=3$, and $k<4$. From the second constraint, $k>3$; therefore $k=4$. From the first constraint, $t<4$; and all possible partitions of $2J_d$, that is 6, into four parts are: 3+1+1+1 and 2+2+1+1.

After partitions are obtained, these partitions are assigned to a_i. Since e_{ii} is the total number of joints belonging to multiple link i and a_i is only the number of joints incident to the other multiple links, $a_i<e_{ii}$. For the sake of avoiding having isomorphic configurations of multiple links, the rule that $a_i<a_j$ is set up, if $e_{ii}=e_{jj}$ and $i>j$. The resultant incident joint sequences H of the above example are: (3,1,1,1), (1,1,3,1), (2,2,1,1), (2,1,2,1), and (1,1,2,2).

9.5.5 Step 5
Construct M_{ul} Matrices

After incident joint sequences are obtained, the permutation group of link elements of M_{ul} matrix can now be determined.

Here, let $S_{ML}=(L_1,L_2,...,L_{Nm})$ be the set of multiple links. The *attribute* of a link is some information about this link, such as e_{ii}, a_i, e_{ij}, ... etc. Up to now, a multiple link has two given attributes, the number of its joints (e_{ii}) and the number of joints incident to the other multiple links (a_i). For example, if $S_{ML}=(L_1,L_2,L_3, L_4)$, the values of e_{ii} are (4,4,3,3), and the values of a_i are (2,2,2,2), then the link-group G_L of the set S_{ML} is:

$$D_L = \{P_{L1}, P_{L2}, P_{L3}, P_{L4}\}$$

where

$$P_{L1} = [L_1][L_2][L_3][L_4]$$

$$P_{L2} = [L_1/L_2][L_3][L_4]$$
$$P_{L3} = [L_1][L_2][L_3/L_4]$$
$$P_{L4} = [L_1/L_2][L_3/L_4]$$

In order to determine the value of e_{ij} for each link i, joint-group D_J that acts on the set of e_{ij} of M_{ul} matrix is used. Each permutation of the link-group induces a corresponding permutation of the joint-group by mapping e_{ij} into e_{pq} if L_i is mapping into L_p and L_j into L_q. For example, the joint-group of the above example is:

$$D_J = \{P_{J1}, P_{J2}, P_{J3}, P_{J4}\}$$

where

$$P_{J1} = [e_{12}][e_{13}][e_{14}][e_{23}][e_{24}][e_{34}]$$
$$P_{J2} = [e_{12}][e_{13}/e_{23}][e_{14}/e_{24}][e_{34}]$$
$$P_{J3} = [e_{12}][e_{13}/e_{14}][e_{23}/e_{24}][e_{34}]$$
$$P_{J4} = [e_{12}][e_{13}/e_{24}][e_{14}/e_{23}][e_{34}]$$

There are three similar classes of D_J here: $\{e_{12}\}$, $\{e_{13},e_{14},e_{23},e_{24}\}$, and $\{e_{34}\}$.

Now, J_d are assigned "1"s to e_{ij} based on joint-group D_J. Unassigned elements are set to "0". In the assignment, the compatibility constraints, Equation (9.10), must be satisfied. A recursive procedure for the assignment problem is provided as follows for this purpose. This procedure, namely ASSIGNMENT, takes four arguments: ESET, GROUP, JD, and PATH. They are initially set equal to the set of elements, permutation group, number of "1"s to be assigned, and an empty list, respectively.

ASSIGNMENT (ESET, GROUP, JD, PATH)
1. if EMPTY (ESET), return;
   EMPTY is a predicate true for its argument that is an empty list. Upon the empty list, the procedure is returned.
2. CLASSLIST <- SIMILAR (ESET, GROUP);
   SIMILAR is a function that separates ESET into some similar classes based on GROUP and orders them into a list.
3. CLASS <- ADDLAST (FIRST(CLASSLIST), CLASSEND);
   The first similar class is selected, and symbol CLASSEND is added, to the last of the class.
4. RSET <- REMOVE (CLASS, ESET);
   REMOVE is a function that removes the elements of CLASS from ESET. The rest of the elements form a set, RSET, for the next recursion.
5. PATHS <- LIST (PATH);
   LIST is a function that produces a list of its argument. A path indicates a sequence of assigned elements.

6. until EMPTY (PATHS), do:
7. begin
8. PATHS <- EXPAND (PATHS, CLASS);
    EXPAND is a function that expands the end node, n_i, of each path PA_i in PATHS to generate a set of new paths. The successors of node n_i are the elements whose order in CLASS is lower than that of n_i.
9. for each path, PA, in the PATHS, do:
10. begin
11. if LAST(PA)=CLASSEND.
    then NGROUP <- MODIFY (GROUP, PA),
    ASSIGNMENT (RSET, NGROUP, REMOVELAST(PA));
    If the path cannot be expanded in the current CLASS, then ASSIGNMENT is called recursively on the rest elements. MODIFY is a function that removes the destroyed permutations from GROUP based on path PA, where a destroyed permutation is one in which the values of elements in the same cycle are distinct.
12. if NOT(MAXIMAL (PA, GROUP)),
    then PATHS <- REMOVE(PA, PATHS), go end;
    MAXIMAL is a predicate true for its first argument, PA, if PA is a maximal path. A maximal path is defined such that the highest order of elements of this path is greater than those of the other paths transformed by permutations of GROUP. If the highest orders are equal, then the second ones are compared, and so on. If PA is not a maximal path, then it must be isomorphic to a maximal path, and be removed from PATHS.
13. if LENGTH (PA)=JD,
    then REMOVE (PA, PATHS),
    if COMPATIBLE (PA), OUTPUT (PA);
    If PA contains J_d assigned elements, then remove PA from PATHS. If PA satisfies compatibility constraints, then it is a result and can be output.
14. end
15. end
16. return

For the example mentioned previously, if the set of joint elements of M_{ul} matrix is $S_{je}=(e_{12},e_{13},e_{14},e_{23},e_{24},e_{34})$, incident joint sequence is $H=(2,2,2,2)$, $J_d=4$, and joint group $D_J=\{P_{J1},P_{J2},P_{J3},P_{J4}\}$, where

$$P_{J1} = [e_{12}][e_{13}][e_{14}][e_{23}][e_{24}][e_{34}]$$
$$P_{J2} = [e_{12}][e_{13}/e_{23}][e_{14}/e_{24}][e_{34}]$$
$$P_{J3} = [e_{12}][e_{13}/e_{14}][e_{23}/e_{24}][e_{34}]$$
$$P_{J4} = [e_{12}][e_{13}/e_{24}][e_{14}/e_{23}][e_{34}]$$

The assignment procedure is executed with its initial arguments

ESET=S_{je}, GROUP=D_J, JD=4, and PATH=nil, where nil denotes an empty list.

The joint-group of the first M_{ul} matrix is the same as the original group, but the joint-group of the second one is modified by removing P_{J2} and P_{J3} from the original group. The corresponding link-group of the second one is reduced as $D_L=\{P_{L1},P_{L4}\}$.

9.5.6 Step 6
Construct M_{ur} Matrices

In Step 5, M_{ul} matrices are constructed. Now, M_{ur} matrices for each M_{ul} matrix are to be constructed. Similar to the process of generating M_{ul} matrices, the process of generating M_{ur} matrices is transformed into the problem of assigning $2N_c$ joints into elements of M_{ur} matrices. There are two differences between this step and Step 5:
1. If the value of the element in M_{ur} matrix is "2", the assignment process must be executed two times. The first time, assign J_2 "2"s to the elements of M_{ur} matrix; and the second time, assign $2N_c-2J_2$ "1"s to the elements. Unassigned elements are set to "0".
2. Assigned elements must satisfy the following compatibility constraints of M_{ur} matrix:

$$\sum_{\substack{j=1 \\ j \neq i}}^{N_m+N_c} e_{ij} = e_{ii}, \qquad i=1,...,N_m$$

$$\sum_{j=1}^{N_m} e_{ij} = 2, \qquad i=N_m+1,....,N_m+N_c$$

Now, let $S_{Lmc}=(L_1,L_2,...,L_n)$ be the set of multiple links and contracted links. The link-group of multiple links with the given M_{ul} matrix is obtained in Step 5. The link-group of contracted links is the set of permutations that transforms the value of e_{ii} into itself. The link-group of the set S_{Lmc} can be obtained by combining link groups of multiple links and contracted links. For example, if $N_L=10$, $F_p=1$, $A_L=[6/2/2/0]$, $A_{LC}=[0/3]$, and

$$M_{CLA} = \begin{bmatrix} 4 & 1 & 1 & 0 & e_{15} & e_{16} & e_{17} \\ 1 & 4 & 0 & 1 & e_{25} & e_{26} & e_{27} \\ 1 & 0 & 3 & 1 & e_{35} & e_{36} & e_{37} \\ 0 & 1 & 1 & 3 & e_{45} & e_{46} & e_{47} \\ e_{51} & e_{52} & e_{53} & e_{54} & -2 & 0 & 0 \\ e_{61} & e_{62} & e_{63} & e_{64} & 0 & -2 & 0 \\ e_{71} & e_{72} & e_{73} & e_{74} & 0 & 0 & -2 \end{bmatrix}$$

then the link-group of multiple links has two permutations: $[L_1][L_2][L_3][L_4]$ and $[L_1/L_2][L_3/L_4]$, that are obtained in Step 5. The link-group of contracted links has six permutations: $[L_5][L_6][L_7]$, $[L_5][L_6/L_7]$, $[L_6][L_5/L_7]$, $[L_7][L_5/L_6]$, $[L_5/L_6/L_7]$, and $[L_5/L_7/L_6]$. The link group D_L of S_{Lmc} has twelve permutations by combining the above two link groups, i.e.,

$$D_L = \{P_{L1}, P_{L2}, P_{L3}, P_{L4}, P_{L5}, P_{L6}, P_{L7}, P_{L8}, P_{L9}, P_{L10}, P_{L11}, P_{L12}\}$$

where

$$P_{L1} = [L_1][L_2][L_3][L_4][L_5][L_6][L_7]$$
$$P_{L2} = [L_1][L_2][L_3][L_4][L_5][L_6/L_7]$$
$$P_{L3} = [L_1][L_2][L_3][L_4][L_6][L_5/L_7]$$
$$P_{L4} = [L_1][L_2][L_3][L_4][L_5/L_6][L_7]$$
$$P_{L5} = [L_1][L_2][L_3][L_4][L_5/L_6/L_7]$$
$$P_{L6} = [L_1][L_2][L_3][L_4][L_5/L_7/L_6]$$
$$P_{L7} = [L_1/L_2][L_3/L_4][L_5][L_6][L_7]$$
$$P_{L8} = [L_1/L_2][L_3/L_4][L_5][L_6/L_7]$$
$$P_{L9} = [L_1/L_2][L_3/L_4][L_6][L_5/L_7]$$
$$P_{L10} = [L_1/L_2][L_3/L_4][L_5/L_6][L_7]$$
$$P_{L11} = [L_1/L_2][L_3/L_4][L_5/L_6/L_7]$$
$$P_{L12} = [L_1/L_2][L_3/L_4][L_5/L_7/L_6]$$

Furthermore, the joint-group D_J of the set $S_{Lmc}=(e_{15}, e_{16}, e_{17}, e_{25}, e_{26}, e_{27}, e_{35}, e_{36}, e_{37}, e_{45}, e_{46}, e_{47})$ is:

$$D_J = \{P_{J1}, P_{J2}, P_{J3}, P_{J4}, P_{J5}, P_{J6}, P_{J7}, P_{J8}, P_{J9}, P_{J10}, P_{J11}, P_{J12}\}$$

where

$$P_{J1} = [e_{15}][e_{16}][e_{17}][e_{25}][e_{26}][e_{27}][e_{35}][e_{36}][e_{37}][e_{45}][e_{46}][e_{47}]$$
$$P_{J2} = [e_{15}][e_{16}/e_{17}][e_{25}][e_{26}/e_{27}][e_{35}][e_{36}/e_{37}][e_{45}][e_{46}/e_{47}]$$
$$P_{J3} = [e_{16}][e_{15}/e_{17}][e_{26}][e_{25}/e_{27}][e_{63}][e_{35}/e_{37}][e_{46}][e_{45}/e_{47}]$$
$$P_{J4} = [e_{15}/e_{16}][e_{17}][e_{25}/e_{26}][e_{27}][e_{35}/e_{36}][e_{37}][e_{45}/e_{46}][e_{47}]$$
$$P_{J5} = [e_{15}/e_{16}/e_{17}][e_{25}/e_{26}/e_{27}][e_{35}/e_{36}/e_{37}][e_{45}/e_{46}/e_{47}]$$
$$P_{J6} = [e_{15}/e_{17}/e_{16}][e_{25}/e_{27}/e_{26}][e_{35}/e_{37}/e_{36}][e_{45}/e_{47}/e_{46}]$$
$$P_{J7} = [e_{15}/e_{25}][e_{16}/e_{26}][e_{17}/e_{27}][e_{35}/e_{45}][e_{36}/e_{46}][e_{37}/e_{47}]$$
$$P_{J8} = [e_{15}/e_{25}][e_{16}/e_{27}][e_{17}/e_{26}][e_{35}/e_{45}][e_{36}/e_{47}][e_{37}/e_{46}]$$
$$P_{J9} = [e_{15}/e_{27}][e_{16}/e_{26}][e_{17}/e_{25}][e_{35}/e_{47}][e_{36}/e_{46}][e_{37}/e_{45}]$$
$$P_{J10} = [e_{15}/e_{26}][e_{16}/e_{25}][e_{17}/e_{27}][e_{35}/e_{46}][e_{36}/e_{45}][e_{37}/e_{47}]$$
$$P_{J11} = [e_{15}/e_{26}/e_{17}/e_{25}/e_{16}/e_{27}][e_{35}/e_{46}/e_{37}/e_{45}/e_{36}/e_{47}]$$
$$P_{J12} = [e_{15}/e_{27}/e_{16}/e_{25}/e_{17}/e_{26}][e_{35}/e_{47}/e_{36}/e_{45}/e_{37}/e_{46}]$$

Finally, the ASSIGNMENT procedure is executed with initial arguments ESET=S_{je}, GROUP=D_J, JD= $2N_c$, and PATH=nil. The resulting M_{ur} matrices are:

$$\begin{bmatrix} 1 & 1 & 0 \\ 1 & 1 & 0 \\ 0 & 0 & 1 \\ 0 & 0 & 1 \end{bmatrix} \quad \begin{bmatrix} 1 & 1 & 0 \\ 1 & 0 & 1 \\ 0 & 1 & 0 \\ 0 & 0 & 1 \end{bmatrix} \quad \begin{bmatrix} 1 & 1 & 0 \\ 1 & 0 & 1 \\ 0 & 0 & 1 \\ 0 & 1 & 0 \end{bmatrix}$$

9.5.7 Step 7
Construct M_{CLA} Matrices

This step is to construct the corresponding M_{CLA} from M_{ur} matrices obtained in Step 6. For the three M_{ur} matrices generated in Step 6, their corresponding M_{CLA} matrices are:

$$\begin{bmatrix} 4 & 1 & 1 & 0 & 1 & 1 & 0 \\ 1 & 4 & 0 & 1 & 1 & 1 & 0 \\ 1 & 0 & 3 & 1 & 0 & 0 & 1 \\ 0 & 1 & 1 & 3 & 0 & 0 & 1 \\ 1 & 1 & 0 & 0 & -2 & 0 & 0 \\ 1 & 1 & 0 & 0 & 0 & -2 & 0 \\ 0 & 0 & 1 & 1 & 0 & 0 & -2 \end{bmatrix} \begin{bmatrix} 4 & 1 & 1 & 0 & 1 & 1 & 0 \\ 1 & 4 & 0 & 1 & 1 & 0 & 1 \\ 1 & 0 & 3 & 1 & 0 & 1 & 0 \\ 0 & 1 & 1 & 3 & 0 & 0 & 1 \\ 1 & 1 & 0 & 0 & -2 & 0 & 0 \\ 1 & 0 & 1 & 0 & 0 & -2 & 0 \\ 0 & 1 & 0 & 1 & 0 & 0 & -2 \end{bmatrix} \begin{bmatrix} 4 & 1 & 1 & 0 & 1 & 1 & 0 \\ 1 & 4 & 0 & 1 & 1 & 0 & 1 \\ 1 & 0 & 3 & 1 & 0 & 0 & 1 \\ 0 & 1 & 1 & 3 & 0 & 1 & 0 \\ 1 & 1 & 0 & 0 & -2 & 0 & 0 \\ 1 & 0 & 0 & 1 & 0 & -2 & 0 \\ 0 & 1 & 1 & 0 & 0 & 0 & -2 \end{bmatrix}$$

Furthermore, each M_{CLA} containing any construction of basic rigid chain should be deleted.

9.5.8 Step 8
Transform M_{CLA} Matrices to Kinematic Chains

The last step is to transform each M_{CLA} obtained in Step 7 into its corresponding kinematic chain in graphic form. Figure 9.9 shows the corresponding kinematic chains of the three M_{CLA} obtained in Step 7.

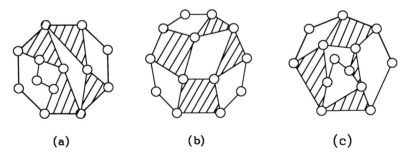

(a) (b) (c)

Figure 9.9 Generated kinematic chains of an example

9.6 Atlas of Kinematic Chains

Table 9.1 summaries the results for the numbers of kinematic chains with six to twelve links and with one to five degrees of freedom.

Table 9.1 Numbers of kinematic chains with 6 to 12 links

N_L \ F	1	2	3	4	5
6	2				
7		4			
8	16		7		
9		40		10	
10	230		98		14
11		839		189	
12	6,862		2,442		354

Furthermore, the atlases of simple kinematic chains are especially useful in designing mechanical devices. Some important atlases of simple kinematic chains are provided here.

For devices with one degree of freedom:
1. There is one (4,4) simple kinematic chain as shown in Figure 9.10.
2. There are two (6,7) simple kinematic chains as shown in Figure 9.11.
3. There are sixteen (8,10) simple kinematic chains as shown in Figure 9.12.

For devices with two degrees of freedom:
1. There is one (5,5) simple kinematic chain as shown in Figure 9.13.
2. There are three (7,8) simple kinematic chains as shown in Figure 9.14.
3. There are forty (9,11) simple kinematic chains as shown in Figure 9.15.

For devices with three degrees of freedom:
1. There is one (6,6) simple kinematic chain as shown in Figure 9.16.
2. There are four (8,9) simple kinematic chains as shown in Figure 9.17.

Figure 9.10   Atlas of the (4,4) kinematic chain

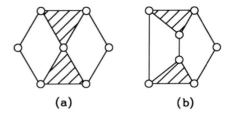

Figure 9.11 Atlas of the (6,7) kinematic chain

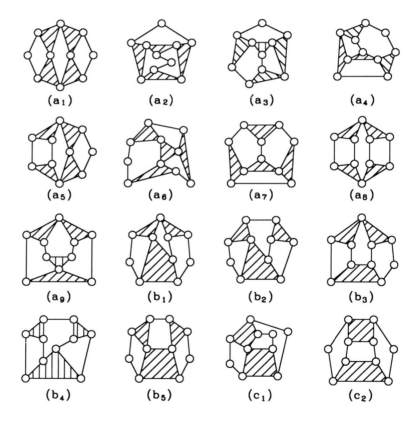

Figure 9.12 Atlas of (8,10) kinematic chains

9.6. Atlas of Kinematic Chains • 155

Figure 9.13 Atlas of the (5,5) kinematic chain

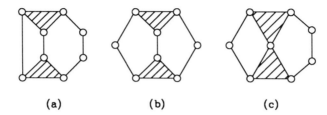

Figure 9.14 Atlas of (7,8) kinematic chains

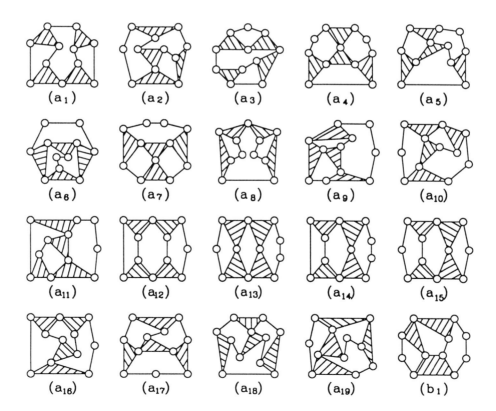

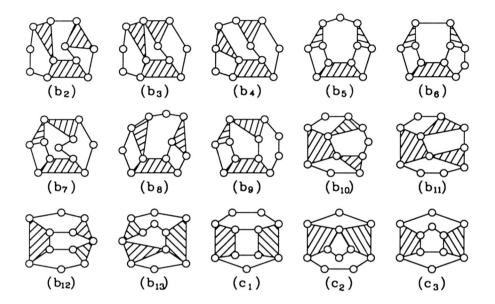

Figure 9.15 Atlas of (9,11) kinematic chains

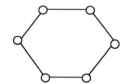

Figure 9.16 Atlas of the (6,6) kinematic chain

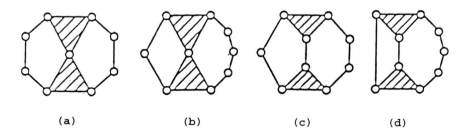

Figure 9.17 Atlas of (8,9) kinematic chains

9.7 Summary

A simple kinematic chain is a kinematic chain with revolute pairs and simple joints only. A basic rigid chain is a rigid chain without any rigid subchain. A kinematic chain with positive degrees of freedom and containing any basic rigid chain is a degenerate kinematic chain that has fewer links.

The atlas of non-isomorphic kinematic chains can also be synthesized based on permutation groups including link-groups, joint-groups, and chain-groups that are defined according to some basic concepts from combinatorial theory. A contract link is a string of binary links in a kinematic chain. Since a contracted link adjacency matrix represents the topological structure of a kinematic chain, the enumeration of kinematic chains can be achieved by constructing all possible contracted link adjacency matrices.

The atlases of simple kinematic chains listed in this chapter provide design engineers the necessary data bank for the generation of all possible topological structures of mechanisms based on the creative design methodology presented in Chapter 6.

Problems

9.1 What are the corresponding link adjacency matrices for the four kinematic chains with eight links and nine joints shown in Figure 9.17?

9.2 What are the corresponding labeled link adjacency matrices for the three kinematic chains with seven links and eight joints shown in Figure 9.14?

9.3 What are the corresponding contracted link adjacency matrices for the two kinematic chains with six links and seven joints shown in Figure 9.11?

9.4 Explain the difference between the topology matrix of a mechanism and the link adjacency matrix of the corresponding kinematic chain of this mechanism.

9.5 Identify the corresponding link-group for the kinematic chain with eight links and ten joints shown in Figure 9.12(a1).

9.6 Identify the corresponding joint-group for the kinematic chain with eight links and ten joints shown in Figure 9.12(b1).

9.7 Identify the corresponding similar classes for the kinematic chain with nine links and eleven joints shown in Figure 9.15(b5).

9.8 Identify the corresponding permutation groups for the kinematic chain with nine links and eleven joints shown in Figure 9.15(c3).

9.9 Enumerate the number of kinematic chains with fourteen links and with one degree of freedom.

References

Hwang, W. M., Computer-aided Structural Synthesis of Planar Kinematic Chains with Multiple Joints, Ph.D. dissertation, Department of Mechanical Engineering, National Cheng Kung University, Tainan, Taiwan, May 1984.

Hwang, W. M. and Yan, H. S., "Atlas of basic rigid chains," Proceedings of the 9th Applied Mechanisms Conference, Session IV-B, No. 1, Kansas City, Missouri, October 28-30, 1985.

Hwang, Y. W., Computer-aided Structural Synthesis of Planar Kinematic Chains with Simple Joints, Master thesis, Department of Mechanical Engineering, National Cheng Kung University, Tainan, Taiwan, May 1986.

Hwang, Y. W., An Expert System for Creative Mechanism Design, Ph.D. dissertation, National Cheng Kung University, Tainan, Taiwan, May 1990.

Yan, H. S., "A methodology for creative mechanism design," *Mechanism and Machine Theory*, Vol. 27, No. 3, 1992, pp. 235-242.

Yan, H. S. and Hwang, Y. W., "Number synthesis of kinematic chains based on permutation groups," *Mathematical and Computer Modeling*, Vol. 13, No. 8, 1990, pp. 29-42.

CHAPTER 10

SPECIALIZATION

The fourth step of the creative design methodology described in Chapter 6 is the process of specialization. This chapter starts with the definition of specialization. Then, the algorithm for the generation of all non-isomorphic specialized devices is presented. Finally, a mathematical expression for counting the number of specialized devices is provided. Based on the algorithm developed in this chapter, designers will be able to obtain the complete atlas of feasible specialized devices with specified types of members and joints, subject to design requirements and constraints.

10.1 Specialized Chains

The process of assigning specific types of members and joints in the available atlas of generalized chains, subject to certain design requirements and constraints, is called *specialization*. A generalized (kinematic) chain after specialization, subject to design requirements is called a *specialized chain*. And a specialized chain subject to design constraints is called a *feasible specialized chain*.

Take the (5,6) generalized chain shown in Figure 10.1(a) for example, and let the design requirements be that the chain should have five revolute pairs (J_R), one cam pair (J_A), and a ground link (K_F). Then, Figures 10.1(b), (c), (d), and (e) show four resulting specialized chains. If the ground link is constrained to be a multiple link, two feasible specialized chains are available, as shown in Figures 10.1(b) and (c).

In what follows, an algorithm for specialization is presented.

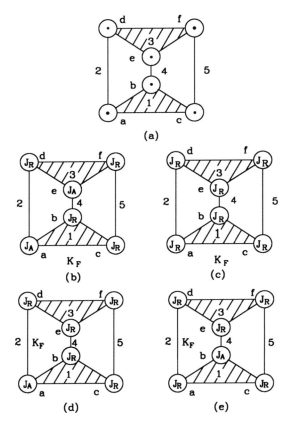

Figure 10.1 A (5,6) generalized chain and its derived (feasible) specialized chains

10.2 Specializing Algorithm

Since each permutation of permutation groups expresses a symmetry of the chain, it is clear that assigning a link or a joint type to similar elements generates isomorphic mechanisms. Therefore, those elements in the same similar class must be ordered such that elements with a higher order have priority of assignment. When a set of assigned elements with a lower order is transformed by a permutation into another set of elements with a higher order, the assignment is abandoned in order not to generate isomorphic mechanisms. After a similar class is assigned, the permutation group is modified by removing the destroyed permutations, whose similar elements in the same cycle are assigned to different types. Based on the new modified group, the rest of the similar classes can be derived to continue the assignment.

10.2. Specializing Algorithm • 161

The procedure for assigning the type of each link or each joint of a generalized (kinematic) chain is as follows:
1. Label each element (link or joint) of a candidate generalized chain.
2. Find the permutation group of the labeled generalized (kinematic) chain from its link adjacency matrix and labeled link adjacency matrix.
3. Based on the permutation group obtained in Step 2, identify the similar classes of unassigned elements.
4. Assign the types to the first similar class by the following substeps:
   (a) Set the order of the elements of the similar class arbitrarily.
   (b) According to the order of these elements, assign a type to one of the elements each time. If a set of assigned elements can be transformed into another set of elements with a higher order, this set of assigned elements is abandoned.
   (c) Repeat step (b) until the assignment of the class is completed.
5. For each result obtained in Step 4, modify the permutation group by removing the destroyed permutations. Then, repeat Steps 3-5 to assign the type to the rest of the similar classes.

[Example 10.1]
Assign the ground link (K_F) to the (6,7) generalized Watt-chain shown in Figure 10.2(a).

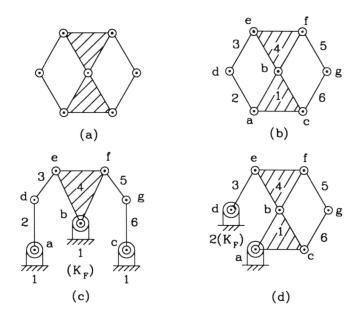

Figure 10.2 The (6,7) Watt-chain and its derived mechanisms

1. All links of this generalized chain are labeled as shown in Figure 10.2(b).
2. The link-group D_L of this generalized chain is:

$$D_L = \{P_{L1}, P_{L2}, P_{L3}, P_{L4}\}$$

where

$P_{L1} = [1][2][3][4][5][6]$
$P_{L2} = [1/4][2/3][5/6]$
$P_{L3} = [1][2/6][3/5][4]$
$P_{L4} = [1/4][2/5][3/6]$

3. The similar classes of this permutation group are {1,4} and {2,3,5,6}.
4. Start with assigning the ground link (K_F) to the first similar class {1,4}, and let the order of the elements of the similar class be {1,4}. Assigning less than one ground link K_F to the order sequence (1,4) has two results as (0,0) and (K_F,0). Here, "K_F" indicates that the element is assigned as the ground link, and "0" indicates that the element is not assigned. Note that the ground link cannot be assigned to link 4, because (0,K_F) can be transformed by P_{L2} (or P_{L4}) into (K_F,0).
5. If the first class is assigned as (K_F,0), the assignment of a ground link is completed and the resulting mechanism is the Watt-II mechanism as shown in Figure 10.2(c). If the first class is assigned as (0,0) with a modified group as the original group, the ground link must be assigned to the other similar classes. Based on the modified group, the similar class is {2,3,5,6}. Assigning the ground link to (2,3,5,6) has one result as (K_F,0,0,0). The other sequences (0,K_F,0,0), (0,0,K_F,0), and (0,0,0,K_F) can be transformed into (K_F,0,0,0) by P_{L2}, P_{L4}, and P_{L3}, respectively. The resulting mechanism is the Watt-I mechanism as shown in Figure 10.2(d).

Therefore, for the Watt-chain shown in Figure 10.2(a), there are two non-isomorphic mechanisms with identified ground link (K_F) available as shown in Figures 10.2(c) and (d).

[Example 10.2]

Assign two prismatic pairs (J_P) and five revolute pairs (J_R) to the joints of the Watt-II mechanism shown in Figure 10.2(c).
1. All joints of this chain are labeled as shown in Figure 10.2(c).
2. Since link permutations P_{L2} and P_{L4} are destroyed in Example 10.1, the corresponding joint permutations P_{J2} and P_{J4} are also destroyed. The joint-group of this mechanism is:

$$D_J = \{P_{J1}, P_{J3}\}$$

where

$P_{J1} = [a][b][c][d][e][f][g]$

$$P_{J3} = [a/c][b][e/f][d/g]$$

3. The similar classes subject to this permutation group are {a,c}, {b}, {e,f}, and {d,g}.
4. Start with assigning J_P pairs to the first similar class, and let the order of the elements of the first similar class be (a,c). The results of assigning less than two J_P pairs to the order pair (a,c) are (J_P,J_P), (J_P,0), and (0,0). Since (0,J_P) can be transformed into (J_P,0) by P_{J3}, it is abandoned.
   (a) If (a,c) is assigned as (J_P,J_P), the assignment is completed, and the resulting mechanism is shown in Figure 10.3(a).
   (b) If (a,c) is assigned as (J_P,0) as shown in Figure 10.4(a), the modified group is {P_{J1}}. However, there is still one J_P pair to be assigned to the other similar classes. Based on this modified group, the similar classes of the unassigned elements are {b}, {e}, {f}, {d}, and {g}. Therefore, the other J_P pairs can be assigned to joints b, e, f, d, or g, respectively. The corresponding mechanisms are shown in Figures 10.3(b)-(f).

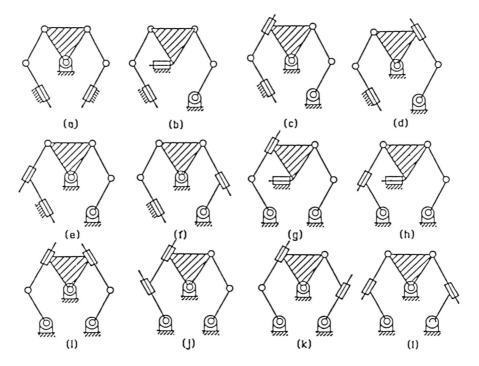

Figure 10.3 Watt-II mechanisms with two prismatic pairs

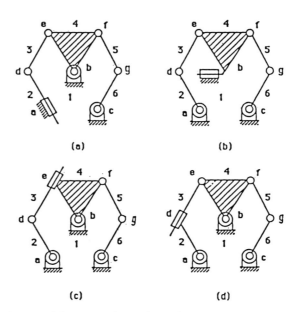

Figure 10.4 Watt-II mechanisms with one prismatic pair

(c) If (a,c) is assigned as (0,0), no permutation is destroyed. There are still two J_P pairs to be assigned to similar classes $\{b\}$, $\{e,f\}$, and $\{d,g\}$. The result of assigning J_P pairs to the first class, $\{b\}$, is (J_P) or (0).

c1. If $(a,c)=(0,0)$ and $(b)=(J_P)$, as shown in Figure 10.4(b), no permutation is destroyed. The rest of the similar classes are $\{e,f\}$ and $\{d,g\}$. The results of assigning J_P pairs to (e,f) are $(J_P,0)$ and $(0,0)$. For the first case, $(e,f)=(J_P,0)$, the resulting mechanism is shown in Figure 10.3(g); and for the second case, $(e,f)=(0,0)$, the other J_P pairs must be assigned to (d,g) as $(J_P,0)$, and the resulting mechanism is shown in Figure 10.3(h).

c2. If $(a,c)=(0,0)$ and $(b)=(0)$, no permutation is destroyed. The other similar classes are $\{e,f\}$ and $\{d,g\}$. The results of assigning J_P pairs to (e,f) are (J_P,J_P), $(J_P,0)$, and $(0,0)$. For the first case, $(e,f)=(J_P,J_P)$, the assignment is completed, and the resulting mechanism is shown in Figure 10.3(i). For the second case, $(e,f)=(J_P,0)$, as shown in Figure 10.4(c), the other classes are $\{d\}$ and $\{g\}$. Therefore, the other J_P pairs can be assigned to $\{d\}$ and $\{g\}$, respectively. The resulting mechanisms are shown in Figures 10.3(j) and (k), respectively. For the third case, $(e,f)=(0,0)$, the other similar class is $\{d,g\}$, and the results of assigning less than two J_P pairs to (d,g) are (J_P,J_P) and $(J_P,0)$. The resulting mechanisms are shown in Figure 10.3(l) and Figure 10.4(d).

Therefore, there are twelve non-isomorphic Watt-II mechanisms with two prismatic pairs as shown in Figure 10.3, and four non-isomorphic Watt-II mechanisms with one prismatic pair as shown in Figure 10.4.

5. Since this chain has seven joints, the remaining five unassigned joints in each case of Figure 10.3 are revolute joints.

In summary, for the Watt-II mechanism shown in Figure 10.2(c), there are twelve non-isomorphic mechanisms with two prismatic pairs and five revolute pairs available, as shown in Figure 10.3.

10.3 Numbers of Specialized Devices

The number of non-isomorphic mechanisms after specialization can be counted by a mathematical formula based on Polya's theory.

Let P_G be a permutation group of a set S. Since each permutation P in P_G can be expressed uniquely as a product of disjoint cycles, the *cycle structure representation of a permutation*, P_{csr}, is defined as:

$$P_{csr} = x_1^{n1} x_2^{n2} \ldots x_k^{nk} \ldots \qquad (10.1)$$

in which x_k is a dummy variable for cycles with a length of k and n_k is the number of cycles with a length of k. For example, the cycle structure representation for the permutation $P=[1][2/6][3/5][4]$, based on Equation (10.1), is $P_{csr} = x_1^2 x_2^2$.

The *cycle index* of a permutation group, P_{ci}, is the summation of the cycle structure representations of all the permutations that constitute the elements of the group divided by the number of permutations, i.e.,

$$P_{ci}(x_1, x_2, x_3, \ldots) = \frac{1}{|P_G|} \sum_{P \in P_G} x_1^{n1} x_2^{n2} x_3^{n3} \qquad (10.2)$$

For the Watt-chain shown in Figure 10.2(b), its cycle index of the link-group is:

$$P_{ci}(x_1, x_2, x_3, \ldots) = (x_1^6 + x_1^2 x_2^2 + 2x_2^3)/4$$

Its cycle index of the joint-group is:

$$P_{ci}(y_1, y_2) = (y_1^7 + y_1^1 y_2^3 + y_1^3 y_2^2 + y_1^1 y_2^3)/4.$$

And its cycle index of the chain-group is:

$$P_{ci}(x_1,x_2;y_1,y_2) = (x_1^6 y_1^7 + x_1^2 x_2^2 y_1^1 y_2^3 + x_2^3 y_1^3 y_2^2 + x_2^3 y_1^1 y_2^3)/4$$

Let S be the set of the links and joints of a candidate generalized chain, S_M be the set of the types (u) of links, S_J be the set of the types (v) of joints, and D_C be the chain-group permutation of S. Then, the *inventory*, I, of a specialized chain or mechanism is:

$$I = P_{ci}\left(\sum u, \sum u^2, \sum u^3, ...; \sum v, \sum v^2, \sum v^3, ...\right) \quad (10.3)$$

The coefficient of each term of Equation (10.3) expresses the number of specialized chains or mechanisms.

[Example 10.3]
Count the number of the Watt-II mechanism, Figure 10.2(c), with two prismatic pairs (J_P) and five revolute pairs (J_R).
Here, $S=\{a,b,c,d,e,f,g\}$, $S_J=\{J_P,J_R\}$, $D_C=D_J=\{P_{J1},P_{J3}\}$, and $P_{ci} = (y_1^7 + y_1^1 y_2^3)/2$. Substituting $y_1=(J_P+J_R)$ and $y_2=(J_P^2+J_R^2)$ into Equation 10.3, inventory I can be expressed as:

$$I = J_P^7 + 4J_P^6 J_R + 12J_P^5 J_R^2 + 19J_P^4 J_R^3 + 19J_P^3 J_R^4 + 12J_P^2 J_R^5 + 4J_P J_R^6 + J_R^7$$

The coefficient of the term $J_P^i J_R^j$ expresses the number of kinematic chains with i prismatic pairs and j revolute pairs. Since the coefficient of the term $J_P^2 J_R^5$ is twelve, there are twelve mechanisms with two prismatic pairs and five revolute pairs. This is consistent with the results of Example 10.2.

[Example 10.4]
Count the number of mechanisms by assigning one ground link (K_F) and two prismatic pairs (J_P) to the Watt-chain shown in Figure 10.2(b).
In this case, based on Equation 10.3, the cycle index P_{ci} of the chain-group of the chain is:

$$(P_{ci}(x_1,x_2;y_1,y_2)) = (x_1^6 y_1^7 + x_1^2 x_2^2 y_1^1 y_2^3 + x_2^3 y_1^3 y_2^2 + x_2^3 y_1^1 y_2^3)/4$$

Substituting $x_1=(K_F+K_L)$, $x_2=(K_F^2+K_L^2)$, $y_1=(J_P+J_R)$, and $y_2=(J_P^2+J_R^2)$ into Equation 10.3, inventory I can be expressed as:

$I =$
$[(K_F+K_L)^6(J_P+J_R)^7+(K_F+K_L)^2(K_F^2+K_L^2)^2(J_P+J_R)(J_P^2+J_R^2)^3$
$+(K_F^2+K_L^2)^3(J_P+J_R)^3(J_P^2+J_R^2)^2+(K_F^2+K_L^2)^3(J_P+J_R)(J_P^2+J_R^2)^3]/4$

The coefficient of the term $K_F K_L^5 J_P^2 J_R^5$ is:

$$\left(\frac{1}{5!2!5!}\right)\left(\frac{d^{13}I}{dK_F d^5 K_L d^2 J_P d^5 J_R}\right) = 33$$

Therefore, there are 33 specialized Watt-mechanisms with identified ground link, two prismatic pairs, and five revolute pairs.

In practical design, the required mechanisms may be subject to various design constraints, such as a type being confined to being assigned to links with certain attributes. And the number of specialized mechanisms subject to certain constraints can also be counted, based on Equation 10.3.

[Example 10.5]
Count the number of specialized mechanisms for the Watt-chain shown in Figure 10.2(a), subject to the following design specifications: (1) the types of links include a ground link (K_F), springs (K_S), and kinematic links (K_L); (2) the ground link must be a multiple link; and (3) the springs must have only two incident joints.

In this case, link set S is separated into two subsets: S_1 and S_2, where S_1 is the set of binary links and S_2 is the set of multiple links. Hence, $S_1=\{2,3,5,6\}$ and $S_2=\{1,4\}$. Cycle index P_{ci} of the link-group must contain two dummy variables x and y to indicate the binary links and the multiple links, respectively, and it is:

$$P_{ci} = (x_1^4 y_1^2 + 2 x_2^2 y_2^1 + x_2^2 y_1^2)/4$$

Now, substituting $x_1=(K_S+K_L)$, $x_2=(K_S^2+K_L^2)$, $y_1=(K_F+K_L)$, and $y_2=(K_F^2+K_L^2)$ into P_{ci}, inventory I can be expressed as:

$I =$
$[(K_S + K_L)^4 (K_F + K_L)^2 + 2(K_S^2 + K_L^2)^2(K_F^2 + K_L^2)$
$+ (K_S^2 + K_L^2)^2 (K_F + K_L)^2]/4$
$=$

$K_F^2(K_S^4 + K_S^3 K_L + 3 K_S^2 K_L^2 + K_S K_L^3 + K_L^4)$
$+ K_F(K_S^4 K_L + 2 K_S^3 K_L^2 + 4 K_S^2 K_L^3 + 2 K_S K_L^4 + K_L^5)$
$(K_S^4 K_L^2 + K_S^3 K_L^3 + 3 K_S^2 K_L^4 + K_S K_L^5 + K_L^6)$

The inventory with a K_F is:

$$K_F(K_S^4 K_L + 2K_S^3 K_L^2 + 4K_S^2 K_L^3 + 2K_S K_L^4 + K_L^5)$$

All of these ten feasible specialized mechanisms are shown in Figure 10.5.

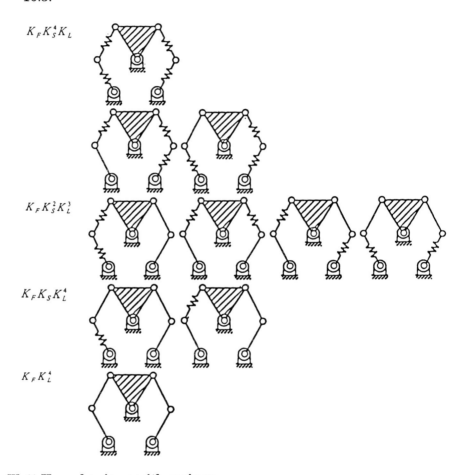

Figure 10.5 Watt-II mechanisms with springs

10.4 Summary

Specialization is one of the major steps of the creative design methodology presented in Chapter 6.

The process of assigning specific types of members and joints in the available atlas of generalized (kinematic) chains subject to certain design requirements and constraints is specialization. A generalized (kinematic) chain after specialization subject to design requirements is a specialized chain. And a specialized chain subject to design constraints is a feasible specialized chain.

The algorithm for the generation of all non-isomorphic specialized chains based on the concepts of permutation groups and similar classes is presented. A mathematical expression for counting the numbers of non-isomorphic specialized chains is provided, based on Polya's theory. And the coefficient of each term of the inventory of a specialized chain expresses the number of the specialized chain.

Based on the algorithm developed in this chapter, designers are able to obtain the complete atlas of (feasible) specialized chains with specified types of members and joints. And a generalized chain may be specialized into various mechanical devices with different topological structures.

Problems

10.1 Count and sketch non-isomorphic specialized mechanisms obtained by assigning one ground link to the atlas of generalized chains with six links and eight joints as shown in Figure 8.14.

10.2 Count and sketch non-isomorphic specialized mechanisms obtained by assigning one ground link to the atlas of kinematic chains with eight links and ten joints as shown in Figure 9.12.

10.3 Count and sketch non-isomorphic specialized mechanisms obtained by assigning one ground link and one spring to the atlas of generalized chains with six links and seven joints as shown in Figure 8.13.

10.4 Count and sketch non-isomorphic specialized mechanisms obtained by assigning two prismatic pairs and five revolute pairs to the kinematic chain with six links and seven joints as shown in Figure 9.7.

10.5 Count and sketch non-isomorphic specialized mechanisms obtained by assigning three prismatic pairs and seven revolute pairs to the kinematic chain with eight links and ten joints as shown in Figure 9.12(a7).

10.6 Count and sketch non-isomorphic specialized mechanisms obtained by assigning one ground link and two prismatic pairs to the kinematic chain with six links and seven joints as shown in Figure 9.7.

10.7 Count and sketch non-isomorphic specialized mechanisms for the generalized chain with six links and seven joints shown in Figure 9.7, subject to the following design specifications: (1) the types of links include one ground link, one actuator, and four kinematic links; (2) the ground link must be a multiple link; and (3) the springs must have only two incident joints.

10.8 Count and sketch non-isomorphic specialized mechanisms for the generalized chain with six links and seven joints shown in Figure 9.1(c), subject to the following design specifications: (1) the types of links include one ground link and five kinematic links; (2) the ground link must be a multiple link; and (3) the types of joints include one cam pair, one prismatic pair, and five revolute pairs.

10.9 How many non-isomorphic specialized mechanisms can be identified for the generalized chain with six links and seven joints shown in Figure 9.1(a), subject to the following design specifications: (1) one of the links is the ground link; and (2) the types of joints can be revolute pairs, rolling pairs, cam pair, and/or gear pairs?

References

Hwang, Y. W., An Expert System for Creative Mechanism Design, Ph.D. dissertation, Department of Mechanical Engineering, National Cheng Kung University, Tainan, Taiwan, May 1990.

Yan, H. S., "A methodology for creative mechanism design," *Mechanism and Machine Theory*, Vol. 27, No. 3, 1992, pp. 235-242.

Yan, H. S. and Hwang, Y. W., "The specialization of mechanisms," *Mechanism and Machine Theory*, Vol. 26, No. 6, 1991, pp. 541-551.

Design Projects

CHAPTER 11

CLAMPING DEVICES

Clamping devices are usually used to hold items for machining operations or to exert great forces for embossing or printing. This chapter synthesizes all possible design configurations that have the same topological characteristics as one abstract clamping device, for the purpose of illustrating the creative design methodology presented in Chapter 6.

11.1 Existing Design

An existing spring loaded clamping device is shown schematically in Figure 11.1, and this concept is patented.

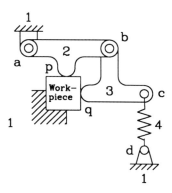

Figure 11.1 A spring loaded clamping device

A design engineer is asked to generate new concepts and to avoid the patent right. After an exhaustive search of available literature and related patents, he concludes that the characteristics of the topological structure of this type of clamping devices are as follows:
1. It is a planar device with four members and six joints.
2. It has a ground link that includes the workpiece (member 1), two clamping links (members 2 and 3), and one spring (member 4).
3. It has four revolute pairs (joints a, b, c, and d) and two direct contacts (joints p and q).
4. It is a structure with -1 degree of freedom.

11.2 Generalization

This spring loaded clamping device is generalized into its corresponding generalized chain, as shown in Figure 11.2, based on the generalizing principles and rules defined in Chapter 7, according to the following steps:

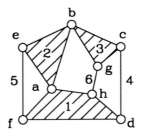

Figure 11.2 Generalized chain of the clamping device

1. The ground link (member 1) is released and generalized into quaternary link 1.
2. Clamping link 2 (member 2) is generalized into ternary link 2.
3. Clamping link 3 (member 3) is generalized into ternary link 3.
4. The spring (member 4), acting as an applied force for the clamping function, is generalized into binary link 4.
5. Direct contact p, that functions as a cam pair, is transformed into binary link 5 with generalized revolute joints e and f at both ends.
6. Direct contact q, that also functions as a cam pair, is transformed into binary link 6 with generalized revolute joints g and h at both ends.

Therefore, it is a generalized chain with six generalized links and eight generalized joints.

11.3 Number Synthesis

Based on the number synthesis of generalized chains described in Chapter 8, and from Figure 8.14, there are nine (6,8) generalized chains, as shown in Figure 11.3.

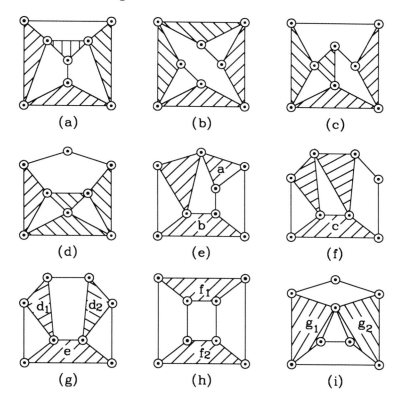

Figure 11.3 Atlas of (6,8) generalized chains of clamping devices

11.4 Specialization

Once the atlas of generalized chains is obtained, all possible specialized chains can be identified according to the following steps:
1. For each generalized chain, identify the ground link and workpiece (member 1) for all possible cases.
2. For each case obtained in Step 1, identify direct contacts.
3. For each case obtained in Step 2, identify clamping links.
4. For each case obtained in Step 3, identify the spring.

These steps are carried out subject to the following design requirements and constraints:

Ground link
1. There must be one ground link to act as the frame and to hold the workpiece.
2. The ground link must be a multiple link to be incident to two direct contacts and to act as the frame.
3. The workpiece that is part of the ground link must be incident to two direct contacts, i.e., there must be at least two binary links adjacent to the workpiece.

Direct contacts
1. There must be two direct contacts, i.e., two binary links.
2. Each direct contact must be incident to the workpiece.

Clamping link
1. There must be at least one clamping link.
2. Each clamping link must be incident to a direct contact.

Spring
1. There must be one spring, i.e., there must be one binary link.
2. The spring must not be incident to the direct contact.

For the nine (6,8) generalized chains shown in Figure 11.3, specialized chains are identified as followers:

Feasible generalized chains
Since the clamping device must have two direct contacts and one spring, a feasible generalized chain should have at least three binary links. Therefore, only those five generalized chains shown in Figures 11.3(e)-(i) are qualified for the process of specialization.

Ground link
Since the ground link must be a multiple link and has at least two binary links adjacent to it, only links a, b, c, d_1, d_2, f_1, f_2, g_1, and g_2 in the five feasible generalized chains shown in Figures 11.3(e)-(i) are qualified to be identified as the ground link. And the ground link can be identified as follows:
1. For the chain shown in Figure 11.3(e), either ternary link a or quaternary link b can be taken as the ground link. Its corresponding generalized devices with an identified ground link are shown in Figures 11.4(a) and (b), respectively.
2. For the chain shown in Figure 11.3(f), only quaternary link c can be taken as the frame. Its corresponding generalized device with an identified ground link is shown in Figure 11.4(c).
3. For the chain shown in Figure 11.3(g) and based on the concept of similar classes, ternary links d_1 (or d_2) and quaternary link e can be

taken as the ground link. Its corresponding generalized devices with an identified ground link are shown in Figures 11.4(d) and (e), respectively.
4. For the chain shown in Figure 11.3(h) and based on the concept of similar classes, either quaternary links f_1 or f_2 can be taken as the ground link. Its corresponding generalized device with an identified ground link is shown in Figure 11.4(f).
5. For the chain shown in Figure 11.3(i) and based on the concept of similar classes, either quaternary links g_1 or g_2 can be taken as the ground link. Its corresponding generalized device with an identified ground link is shown in Figure 11.4(g).

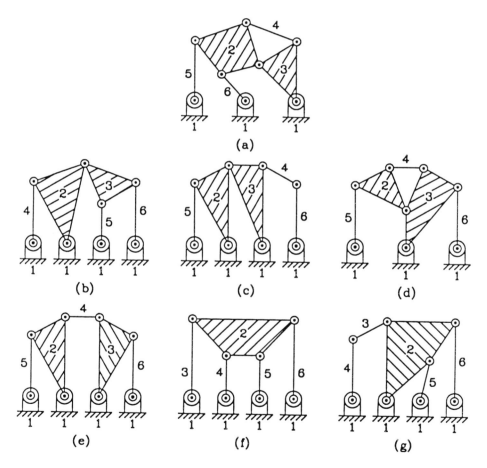

Figure 11.4 Specialized clamping devices with identified ground link

Therefore, there are seven generalized devices with an identified ground link available, as shown in Figure 11.4.

Direct contacts

Since a direct contact is incident to the workpiece (i.e., the ground link) and is generalized into a binary link with generalized revolute joints at both ends, the two direct contacts can be identified as follows:
1. For the generalized device shown in Figure 11.4(a), binary links 5 and 6 are the two direct contacts, as shown in Figure 11.5(a).
2. For the generalized device shown in Figure 11.4(b) and based on the concept of similar classes for links 5 and 6, two cases are available as follows:
   (a) Binary links 4 and 5 are the two direct contacts, as shown in Figure 11.5(b).
   (b) Binary links 5 and 6 are the two direct contacts, as shown in Figure 11.5(c).
3. For the generalized device shown in Figure 11.4(c), binary links 5 and 6 are the two direct contacts, as shown in Figure 11.5(d).
4. For the generalized device shown in Figure 11.4(d), binary links 5 and 6 are the two direct contacts, as shown in Figure 11.5(e).
5. For the generalized device shown in Figure 11.4(e), binary links 5 and 6 are the two direct contacts, as shown in Figure 11.5(f).
6. For the generalized device shown in Figure 11.4(f) and based on the concept of similar classes for links 3, 4, 5, and 6, either two of links 3, 4, 5, and 6 can be identified as the two direct contacts. Here, links 5 and 6 are identified as the two clamping pairs, as shown in Figure 11.5(g).
7. For the generalized device shown in Figure 11.4(g) and based on the concept of similar classes for links 5 and 6, two cases are available as follows:
   (a) Binary links 4 and 5 are the two direct contacts, as shown in Figure 11.5(h).
   (b) Binary links 5 and 6 are the two direct contacts, as shown in Figure 11.5(i).

Therefore, there are nine devices with the ground link and two direct contacts identified, as shown in Figures 11.5(a)-(i)

Clamping link

Since there must be at least one clamping link, and each clamping link must be incident to a direct contact, clamping link(s) can be identified as follows:
1. For the specialized device shown in Figure 11.5(a), link 2 is a clamping link.
2. For the specialized device shown in Figure 11.5(b), links 2 and 3 are clamping links.
3. For the specialized device shown in Figure 11.5(c), link 3 is a clamping link.

11.4. Specialization • 179

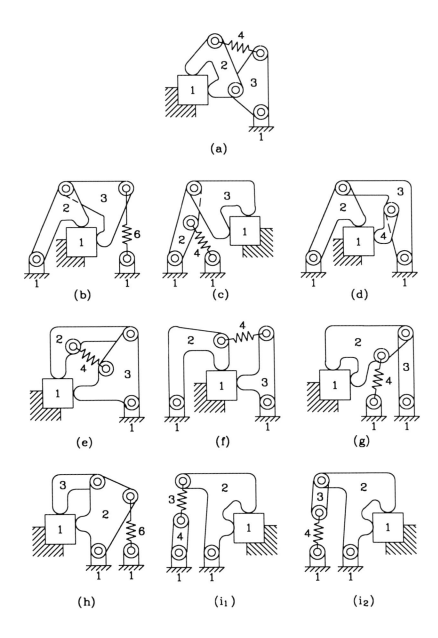

Figure 11.5 Specialized clamping devices with
identified ground link, direct contacts, clamping link(s), and spring

4. For the specialized device shown in Figure 11.5(d), links 2 and 4 are clamping links.
5. For the specialized device shown in Figure 11.5(e), links 2 and 3 are clamping links.
6. For the specialized device shown in Figure 11.5(f), links 2 and 3 are clamping links.
7. For the specialized device shown in Figure 11.5(g), link 2 is a clamping link.
8. For the specialized device shown in Figure 11.5(h), links 2 and 3 are clamping links.
9. For the specialized device shown in Figure 11.5(i), link 2 is a clamping link.

Therefore, they are also nine devices with the ground link, two direct contacts, and one or two clamping links identified, as shown in Figures 11.5(a)-(i).

Spring

Since the spring is generalized into a binary link with generalized revolute joints at both ends and it cannot be adjacent to a direct contact, the spring can be identified as follows:

1. For the device shown in Figure 11.4(a), since binary links 5 and 6 are identified as the direct contacts, binary link 4 is the spring, as shown in Figure 11.5(a).
2. For the device shown in Figure 11.4(b), two cases are available as follows:
   (a) Since binary links 4 and 5 are identified as the direct contacts, binary link 6 is the spring, as shown in Figure 11.5(b).
   (b) Since binary links 5 and 6 are identified as the direct contacts, binary link 4 is the spring, as shown in Figure 11.5(c).
3. For the device shown in Figure 11.4(c), since binary links 5 and 6 are identified as the direct contacts, and binary link 4 is identified as the clamping link, there is no binary link available to be identified as a spring. Therefore, this device is not feasible, Figure 11.5(d).
4. For the device shown in Figure 11.4(d), since binary links 5 and 6 are identified as the direct contacts, binary link 4 is the spring, as shown in Figure 11.5(e).
5. For the device shown in Figure 11.4(e), since binary links 5 and 6 are identified as the direct contacts, binary link 4 is the spring, as shown in Figure 11.5(f).
5. For the device shown in Figure 11.4(f) and based on the concept of similar classes for links 3, 4, 5, and 6, since binary links 5 and 6 are identified as the direct contacts, either one of binary links 3 or 4 can be identified as the spring. Here, link 4 is identified as the spring, as shown in Figure 11.5(g).

7. For the device shown in Figure 11.4(g), two cases are available as follows:
   (a) Since binary links 4 and 5 are identified as the direct contacts and binary link 3 is a clamping member, binary link 6 is the spring, as shown in Figure 11.5(h).
   (b) For the case such that binary link 2 is the clamping link, binary link 3 or 4 is the spring, as shown in Figure 11.5(i_1) or (i_2).

Therefore, there are eight clamping devices synthesized, as shown in Figures 11.5(a)-(c) and (e)-(i_1 or i_2).

In order to make the two devices shown in Figures 11.5(i_1) and (i_2) workable, members 3 and 4 must be in-line. This means that the connecting member (member 3 or 4) is redundant in each case and the device degenerates into three members. Therefore, these two devices are not feasible. And there are only seven feasible specialized clamping devices available, as shown in Figures 11.5(a)-(c) and (e)-(h).

11.5 Particularization

By redrawing these seven feasible devices, as shown in Figures 11.5(a)-(c) and (e)-(h), to make them look right, the atlas of clamping devices is obtained as shown in Figures 11.6(a), (b), (c), (d), (e), (f), and (g), respectively.

11.6 Atlas of New Clamping Devices

The device shown in Figure 11.6(b) is the original clamping device. Therefore, the other six clamping devices shown in Figure 11.6 can be claimed as new design concepts.

11.7 Remarks

If binary link 3 shown in Figure 11.6(f) is further specialized into a direct contact, another new design concept, as shown in Figure 11.7, is available. In order to make use of these design concepts, force analysis should be carried to find the optimum location for the spring.

Problems

11.1 For each clamping device shown in Figure 11.6, identify its corresponding topology matrix.

182 ▪ 11. Clamping Devices

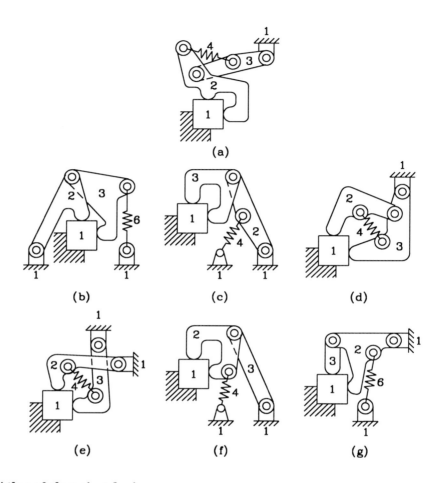

Figure 11.6 Atlas of clamping devices

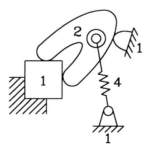

Figure 11.7 A clamping device with three direct contacts

11.2 For each clamping device shown in Figure 11.6, identify its corresponding generalized chain.
11.3 Conclude the topological characteristics of the clamping devices shown in Figure 11.6 based on the technique of attribute listing.
11.4 Count the number of specialized clamping devices with identified ground link, for the nine (6,8) generalized chains shown in Figure 11.3, if no design constraints are imposed on the ground link.
11.5 Count the number of specialized clamping devices with identified ground link, for the nine (6,8) generalized chains shown in Figure 11.3, with the required design constraints.
11.6 Synthesize all possible design configurations that have the same topological characteristics as the clamping device shown in Figure 11.6(a).
11.7 Synthesize all possible design configurations that have the same topological characteristics as the clamping device shown in Figure 11.7.
11.8 Synthesize all possible design configurations that have the same topological characteristics as the clamping device shown in Figure 11.1, if the design constraint that each clamping link should have only one direct contact is further required.
11.9 Synthesize the topological structure of some clamping devices that have the same clamping function as the one shown in Figure 11.1, if the number of members of the devices is required to be larger than four.

References

Hall, A. S. Jr., Generalized Linkages Forms of Mechanical Devices, ME261 class notes, Purdue University, West Lafayette, Indiana, spring 1978.

Yan, H. S., "A methodology for creative mechanism design," *Mechanism and Machine Theory*, Vol. 27, No. 3, 1992, pp. 235-242.

Yan, H. S. and Hwang, Y. W., "The generalization of mechanical devices," *Journal of the Chinese Society of Mechanical Engineers (Taiwan)*, Vol. 9, No. 4, 1988, pp. 283-293.

CHAPTER 12

MOTORCROSS SUSPENSION MECHANISMS

Suspension systems of motorcycles are used to absorb any road shocks that result from the wheels hitting holes or bumps in the road. The objective of single shock rear suspensions of motorcrosses is to keep the real wheel on the ground as much as possible, while providing a reasonably comfortable ride. This chapter synthesizes all possible design configurations of motorcross single shock rear suspensions that have the same topological characteristics as some existing designs, based on the creative design methodology presented in Chapter 6.

12.1 Existing Designs

In general, the rear wheel suspension mechanism of a motorcycle is either a telescope type or a four-bar linkage. However, such a design cannot provide a large wheel travel with variable leverage ratio. Therefore, the concept of six-bar linkages is adopted for the rear suspension of motorcrosses. Honda CR250R pro-link, as shown in Figure 12.1(a), is such a design.

An engineer is assigned the task of coming up with new designs that have the same topological characteristics as the Honda pro-link. He starts the work according to the rational problem solving approach as described in Chapter 4, and he identifies two more products based on the concept of six-bar linkages. They are Suzuki RM250X full-floater, Figure 12.2(a); and Kawasaki KX250 uni-trak, Figure 12.3(a). By studying these existing designs, the engineer concludes the characteristics of their topological structures are as follows:

186 ▪ 12. Motorcross Suspension Mechanisms

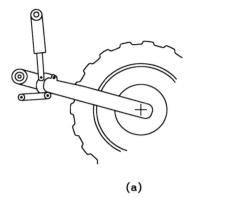
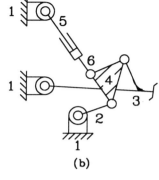

(a)   (b)

Figure 12.1   Honda pro-link suspension

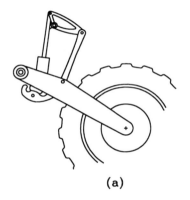

(a)   (b)

Figure 12.2   Suzuki full-floater suspension

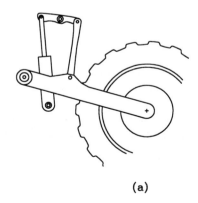
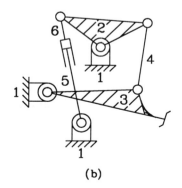

(a)   (b)

Figure 12.3   Kawasaki uni-trak suspension

1. They are planar six-bar mechanisms with seven joints.
2. They have a ground link (member 1, K_F), a pivot arm (member 2, K_{Lp}), a swing arm (member 3, K_{Ls}), a connecting link (member 4, K_{Lc}), and a shock absorber (K_T) consisting of a piston (member 5, K_I) and a cylinder (member 6, K_Y).
3. They have six revolute pairs (joints a, b, c, d, e, and f; J_R) and one prismatic pair (joint g; J_P).
4. They are mechanisms with one degree of freedom.

Figures 12.1(b), 12.2(b), and 12.3(b) are the corresponding schematic representations of the three existing designs shown in Figures 12.1(a), 12.2(a), and 12.3(a), respectively. The topology matrix (M_T), as defined in Chapter 2, of the Suzuki design shown in Figure 12.2 is:

$$M_T = \begin{bmatrix} K_F & J_R & J_R & 0 & 0 & 0 \\ a & K_{Lp} & 0 & J_R & J_R & 0 \\ b & 0 & K_{Ls} & J_R & 0 & J_R \\ 0 & c & e & K_{Lc} & 0 & 0 \\ 0 & d & 0 & 0 & K_I & J_P \\ 0 & 0 & f & 0 & g & K_Y \end{bmatrix}$$

12.2 Generalization

The Suzuki full-floater shown in Figure 12.2 is selected arbitrarily as the original mechanism for the process of generalization. The corresponding generalized chain as shown in Figure 12.4 is obtained, based on the generalizing rules defined in Chapter 7, according to the following procedure:
1. The ground link (member 1) is released and generalized into binary link 1.
2. The pivot arm (member 2) is generalized into ternary link 2.

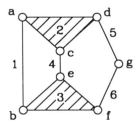

Figure 12.4 Generalized (kinematic) chain of Suzuki full-floater suspension

3. The swing arm (member 3) is generalized into ternary link 3.
4. The connecting link (member 4) is generalized into binary link 4.
5. The piston (member 5) and the cylinder (member 6) of the shock absorber are generalized into a dyad (binary links 5 and 6).
6. The prismatic pair (joint g) is generalized into revolute joint g.

Therefore, the generalized chain has six generalized links and seven generalized revolute joints, and it is a (6,7) generalized kinematic chain.

12.3 Number Synthesis

Based on the number synthesis of kinematic chains described in Chapter 9, and from Figure 9.12, there are two (6,7) kinematic chains, as shown in Figure 12.5.

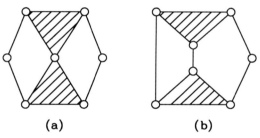

(a)   (b)

Figure 12.5 Atlas of (6,7) kinematic chains of motorcross suspensions

12.4 Specialization

Once the atlas of kinematic chains is obtained, all possible specialized chains can be identified, through the following steps:
1. For each kinematic chain, identify the ground link for all possible cases.
2. For each case obtained in Step 1, identify the shock absorber.
3. For each case obtained in Step 2, identify the swing arm.
4. For each case obtained in Step 3, identify the pivot arm.
5. For each case obtained in Step 4, identify the connecting link.

These steps are carried out subject to the following design requirements and constraints, based on the concluded characteristics of the motorcross suspension mechanisms:
1. There must be a ground link as the frame.
2. There must be a shock absorber.
3. There must be a swing arm.
4. There must be a pivot arm.

5. There must be a connecting link.
6. The ground link, the shock absorber, and the swing arm must be distinct members.

For the two (6,7) kinematic chains shown in Figure 12.5, all possible feasible specialized chains are identified as follows:

Ground link (K_F)

Since there must be a link as the frame and there is no constraint on the assignment of the ground link, the ground link can be identified as follows:

1. For the kinematic chain shown in Figure 12.5(a) and based on the concept of similar classes, the assignment of the ground link generates two non-isomorphic results, as shown in Figures 12.6(a) and (b).
2. For the kinematic chain shown in Figure 12.5(b) and based on the concept of similar classes, the assignment of the ground link generates three non-isomorphic results, as shown in Figures 12.6(c)-(e).

Therefore, five specialized chains with one identified ground link are available, as shown in Figure 12.6.

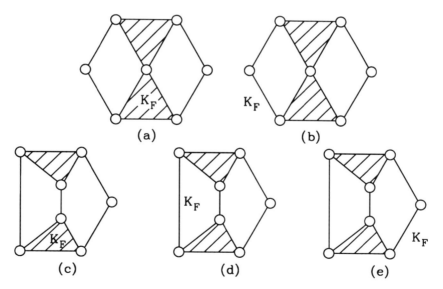

Figure 12.6 Specialized chains with identified ground link

Shock absorber (K_T)

Since there must be a shock absorber consisting of a pair of binary links (dyad) and either the piston or the cylinder cannot be fixed the frame, the shock absorber can be identified as follows:

1. For the case shown in Figure 12.6(a), either one of the two dyads can be assigned as the shock absorber, Figure 12.7(a).
2. For the case shown in Figure 12.6(b), the available dyad is assigned as the shock absorber, Figure 12.7(b).
3. For the case shown in Figure 12.6(c), the dyad is assigned as the shock absorber, Figure 12.7(c).
4. For the case shown in Figure 12.6(d), the dyad is assigned as the shock absorber, Figure 12.7(d).
5. For the case shown in Figure 12.6(e), no dyad is available to be assigned as the shock absorber.

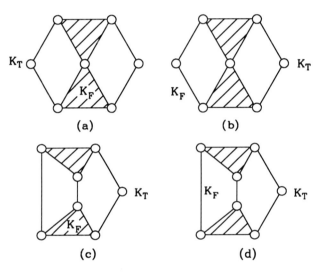

Figure 12.7 **Specialized chains with identified ground link and shock absorber**

Therefore, four specialized chains with identified ground link and shock absorber are available, as shown in Figure 12.7.

Swing arm (K_{Ls})

Since there must be a swing arm, and the ground link and the shock absorber cannot be assigned as the swing arm, the swing arm can be identified as follows:

1. For the case shown in Figure 12.7(a), the assignment of the swing arm generates three results, as shown in Figures 12.8 (a)-(c).
2. For the case shown in Figure 12.7(b), the assignment of the swing arm generates three results, as shown in Figures 12.8 (d)-(f).
3. For the case shown in Figure 12.7(c) and based on the concept of similar classes, the assignment of the swing arm generates two non-isomorphic results, as shown in Figures 12.8 (g)-(h).

4. For the case shown in Figure 12.7(d) and based on the concept of similar classes, the assignment of the swing arm generates two non-isomorphic results, as shown in Figures 12.8 (i)-(j).

Therefore, ten specialized chains with identified ground link, shock absorber, and swing arm are available, as shown in Figure 12.8.

Pivot arm (K_{Lp})

Since there must be a pivot arm that is adjacent to the ground link, the pivot arm can be identified as follows:
1. For the case shown in Figure 12.8(a), binary link 2 is the pivot arm.
2. For the case shown in Figure 12.8(b), ternary link 2 is the pivot arm.
3. For the case shown in Figure 12.8(c), binary link 2 is the pivot arm.
4. For the case shown in Figure 12.8(d), binary link 2 is the pivot arm.
5. For the case shown in Figure 12.8(e), binary link 2 is the pivot arm.
6. For the case shown in Figure 12.8(f), ternary link 2 is the pivot arm.
7. For the case shown in Figure 12.8(g) and based on the concept of similar classes for binary links 2 and 3, either link 2 or link 3 can be the pivot arm. Here, link 2 is assigned as the pivot arm.
8. For the case shown in Figure 12.8(h), binary link 2 is the pivot arm.
9. For the case shown in Figure 12.8(i), ternary link 2 is the pivot arm.
10. For the case shown in Figure 12.8(j) and based on the concept of similar classes for ternary links 2 and 3, either link 2 or link 3 can be the pivot arm. Here, link 2 is assigned as the pivot arm.

Therefore, ten specialized chains with identified ground link, shock absorber, swing arm, and pivot arm are available, as shown in Figure 12.8.

Connecting link (K_{Lc})

The remaining member in each of Figures 12.8(a)-(j) is the connecting link.

In conclusion, there are ten feasible specialized chains available, as shown in Figure 12.8.

12.5 Particularization

The next step of the creative design methodology is to particularize each feasible specialized chain by applying the generalizing rules in reverse to obtain the corresponding schematic diagram of the motorcross suspension mechanisms. For those ten feasible specialized chains shown in Figures 12.8(a)~(j), their corresponding mechanisms are shown in Figures 12.9(a)~(j), respectively.

12.6 Atlas of New Motorcross Suspension Mechanisms

It is interesting to note that Figure 12.9(b) is the Kawasaki uni-track as shown in Figure 12.3, Figure 12.9(h) is the Honda pro-link as shown in Figure 12.1, and Figure 12.9(i) is the Suzuki full-floater as shown in Figure 12.2. Therefore, the remaining seven design concepts as shown in Figures 12.9(a), (c), (d), (e), (f), (g), and (j) are new for motorcross rear suspension mechanisms.

12.7 Remarks

Design requirements are firm to guarantee the results with desired topological characteristics. However, design constraints can be flexible. They are normally identified based on engineering reality and the designers' decisions. The creative techniques provided in Chapter 5 are very helpful for concluding design constraints. Different design constraints result in different atlases of feasible specialized chains. For example, if the design constraint that the ground link, the shock absorber, and the swing arm must be distinct is released, many more design configurations can be synthesized.

Problems

12.1 For each motorcross rear suspension shown in Figure 12.9, identify its corresponding topology matrix.

12.2 For each motorcross rear suspension shown in Figure 12.9, identify its corresponding generalized chain.

12.3 Conduct a brainstorming session to define design requirements and constraints for motorcross rear suspensions.

12.4 Count the number of specialized chains, for motorcross rear suspensions with identified ground link, for the two (6,7) generalized kinematic chains shown in Figure 12.5.

12.5 Count the number of specialized chains, for motorcross rear suspensions with identified ground link and shock absorber, for the two (6,7) generalized kinematic chains shown in Figure 12.5.

12.6 Synthesize all possible design configurations that have the same topological characteristics as the motorcross rear suspension shown in Figure 12.3.

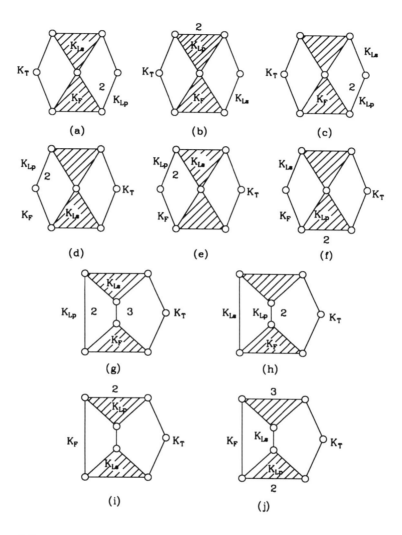

Figure 12.8 Atlas of (feasible) specialized chains for motorcross suspensions

    12.7 Synthesize all possible design configurations that have the same topological characteristics as the motorcross rear suspension shown in Figure 12.1, if the constraint that the swing arm must be adjacent to the ground link is required.

    12.8 Synthesize all possible design configurations that have the same topological characteristics as the motorcross rear suspension shown in Figure 12.1, if the design constraint that the ground link, the shock absorber, and the swing arm must be distinct is released.

194 ■ 12. Motorcross Suspension Mechanisms

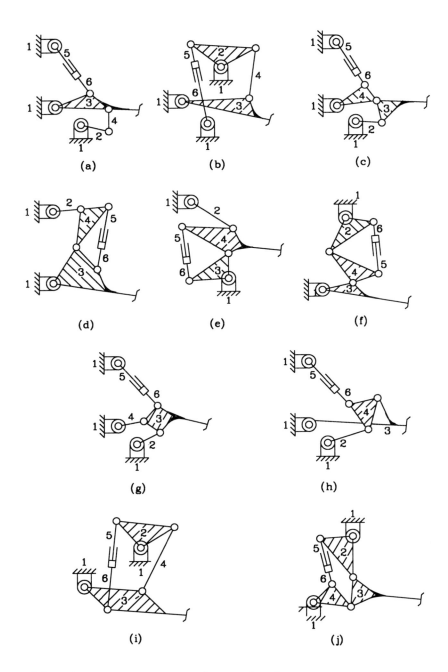

Figure 12.9 Atlas of motorcross suspension mechanisms

12.9 Synthesize some design configurations of motorcross rear suspension mechanisms that have the same topological characteristics as the one shown in Figure 12.2, except the number of members of the design is eight instead of six.

References

Yan, H. S., "A methodology for creative mechanism design," *Mechanism and Machine Theory*, Vol. 27, No. 3, 1992, pp. 235-242.

Yan, H. S. and Chen, J. J., "Creative design of a wheel damping mechanism," *Mechanism and Machine Theory*, Vol. 20, No. 6, 1985, pp. 597-600.

CHAPTER 13

INFINITELY VARIABLE TRANSMISSIONS

Planetary gear trains are mechanisms in which at least one member is required to rotate about its own axis and at the same time to revolve about another axis to provide constant velocity ratios. They are used in various transmission systems due to the advantages of lightweight, compact, and having a high gear ratio. One application of planetary gear trains is in infinitely variable transmissions. This chapter synthesizes all possible design configurations of planetary gear trains for infinitely variable transmissions that have the desired topological characteristics as an existing design, based on the creative design methodology presented in Chapter 6.

13.1 Existing Design

In general, an *infinitely variable transmission* (IVT) consists of a continuous variable unit (CVU) and a planetary gear train (PGT) with two degrees of freedom. Since an IVT has the kinematic property of zero velocity ratio, it can be used as the power train for exercising machines. Figure 13.1 shows the schematic drawing of such a design. It consists of an input-coupled CVU and a five-bar PGT with two degrees of freedom.

An important characteristic that governs the performance of infinitely variable transmissions is the topological structure of the planetary gear train. Figure 13.2 shows the schematic representation of this planetary mechanism. It has one carrier (member 2, K_{Lc}), adjacent to the ground link (member 1, K_F) with a revolute pair (joint a, J_R) and to a planet gear (member 3, K_{Gp}) with a revolute pair (joint d,

J_R), as the output. It has a sun gear (member 4, K_{Gs}), adjacent to the ground link with a revolute pair (joint b, J_R) and meshing with the planet gear with a gear pair (joint e, J_G), as one input (input I). It has a ring gear (member 5, K_{Gr}), adjacent to the ground link with a revolute pair (joint c, J_R) and meshing with the planet gear with a gear pair (joint f, J_G) as another input (input II). The topology matrix (M_T) of this mechanism, as defined in Chapter 2, is:

$$M_T = \begin{bmatrix} K_F & J_R & 0 & J_R & J_R \\ a & K_{Lc} & J_R & 0 & 0 \\ 0 & d & K_{Gp} & J_G & J_G \\ b & 0 & e & K_{Gs} & 0 \\ c & 0 & f & 0 & K_{Gr} \end{bmatrix}$$

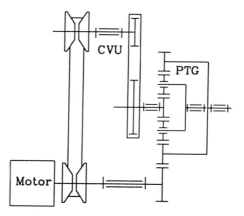

Figure 13.1 **An infinitely variable transmission**

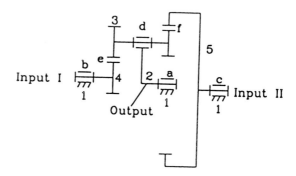

Figure 13.2 **Planetary gear train of the infinitely variable transmission**

13.1. Existing Design

By studying this design and based on the techniques of rational problem solving as presented in Chapter 4, it is found that the planetary gear train for the infinitely variable transmission has the following topological characteristics:

1. It has five members, consisting of at least one ground link, one carrier, one planet gear, one sun gear, and one ring gear.
2. It has six joints that contains four revolute pairs and two gear pairs.
3. It is a reverted gear train.
4. The sun gear and the ring gear are the two inputs, and the carrier is the output.
5. All gears are spur gears.
6. It has two degrees of freedom.

Based on Equation (2.1) for mobility analysis, the following expression is true for a planetary gear train with two degrees of freedom ($F_P=2$), N_L members, N_{JR} revolute pairs, and N_{JG} gear pairs:

$$2 N_{JR} + N_{JG} - 3N_L + 5 = 0 \qquad (13.1)$$

It is obvious that the following expression is also true for a planetary gear train with N_J joints:

$$N_{JR} + N_{JG} - N_J = 0 \qquad (13.2)$$

Furthermore, every link in a geared kinematic chain has at least one revolute pair. By removing all gear pairs from the geared kinematic chain, it forms a tree. It is well known from graph theory that a tree with p nodes (N_L links) contains $q-1$ edges (N_J-1 joints). Therefore, the following expression for a planetary gear train is true:

$$N_{JR} - N_L + 1 = 0 \qquad (13.3)$$

By solving Equations (13.1), (13.2), and (13.3) for the numbers of joints (N_J, N_{JR}, and N_{JG}) in terms of the number of links (N_L), the following relations for planetary gear trains with two degrees of freedom are concluded:

$$N_J = 2N_L - 4 \qquad (13.4)$$

$$N_{JR} = N_L - 1 \qquad (13.5)$$

$$N_{JG} = N_L - 3 \qquad (13.6)$$

Equations (13.4), (13.5), and (13.6) indicate that for a planetary gear train with two degrees of freedom and with five members, it always has six joints, consisting of four revolute pairs and two gear pairs.

13.2 Generalization

Once an existing design is identified and the characteristics of the planetary gear train for infinitely variable transmissions are obtained, the next step of the creative design methodology is to transform this design into its corresponding generalized chain based on the generalizing rules defined in Chapter 7. The generalization of the planetary gear train shown in Figure 13.2 is carried out according to the following steps:
1. The ground link (member 1) is released and generalized into ternary link 1.
2. The carrier (member 2) is generalized into binary link 2.
3. The planet gear (member 3) is generalized into ternary link 3.
4. The sun gear (member 4) is generalized into binary link 4.
5. The ring gear (member 5) is generalized into binary link 5.
6. The revolute pairs between the ground link and the carrier, the ground link and the sun gear, the ground link and the ring gear, and the carrier and the planet gear are generalized into generalized revolute joints a, b, c, and d, respectively.
7. The gear pairs between the planet gear and the sun gear, and the planet gear and the ring gear are generalized into generalized gear joints e and f, respectively.

Therefore, the corresponding generalized chain of the planetary gear train shown in Figure 13.2 has five generalized links and six generalized joints, as shown in Figure 13.3.

13.3 Number Synthesis

Once the generalized chain of an existing design is obtained, the next step of the creative design methodology is to have the atlas of generalized chains with the required numbers of links and joints, based on the algorithm of number synthesis of generalized chains as described in Chapter 8. From the point of view of applications, designers can simply identify the desired chains from various atlases as provided in Chapter 8. For example, from Figure 8.11, two (5,6) generalized chains are available, as shown in Figure 13.4.

13.4 Design Requirements and Constraints

Based on the topological structures of the available existing design and the creative techniques provided in Chapter 5, design requirements and constraints of planetary gear trains for infinitely variable transmissions are carefully concluded as follows:

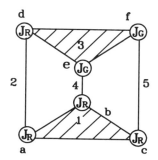

Figure 13.3 Generalized chain of the planetary gear train

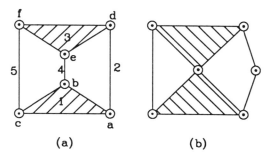

Figure 13.4 Atlas of (5,6) generalized chains of infinitely variable transmissions

Ground link
1. One of the links in each generalized chain must be the ground link.
2. A ground link must be a multiple link in order to have two input members and one output member.
3. A ground link must not be included in a three-bar loop in the chain, because a planetary transmission is a reverted gear train.

Planet gear
1. There must be at least one planet gear.
2. Any link that is not adjacent to the ground link is a planet gear.
3. A planet gear that is not adjacent to another planet gear must not be included in a three-bar loop.
4. A planet gear must be a multiple link including at least one gear pair and a revolute pair incident to the carrier, in order to avoid degeneration.

Carrier
1. There must be a carrier corresponding to each planet gear.
2. A carrier must be adjacent to both the planet gear and the ground link.

3. Two or more planet gears in series must share a common carrier, in order to maintain the center distance between them.

Sun gear
1. There must be at least one sun gear.
2. Any link that is adjacent to the ground link and is not a carrier is a sun gear.

Revolute pair
1. There must be N_L-1 revolute pairs.
2. Every link must have at least one revolute pair.
3. Any joint incident to the ground link must be a revolute pair.
4. The common incident joint of a planet gear and a carrier must be a revolute pair.
5. A planet gear can have only one revolute gear.
6. There can be no loop formed exclusively by revolute pairs.

Gear pair
1. There must be N_L-3 gear pairs.
2. Any joint incident to both a planet gear and a sun gear must be a gear pair.
3. There can be no three-bar loop formed exclusively by gear pairs.

Design requirements and constraints can be flexible, and can be varied for different cases and expectations. If a design engineer thinks that the two (5,6) generalized chains shown in Figure 13.4 might have limited room for generating numerous potential design concepts, he can consider planetary gear trains with two degrees of freedom and with six members, instead of five, for the infinitely variable transmissions. In such a case, i.e., for a planetary gear train with two degrees of freedom and with six members, based on Equations (13.4), (13.5), and (13.6), it always has eight joints, consisting of five revolute pairs and three gear pairs. And from Figure 8.14, there are nine (6,8) generalized chains available for specialization, as shown in Figure 13.5.

13.5 Specialization

The next step of the creative design methodology is to identify the corresponding feasible specialized chains from the available atlas of generalized chains, subject to the necessary design requirements and constraints, according to the following steps:
1. For each generalized chain, identify the ground link for all possible cases.

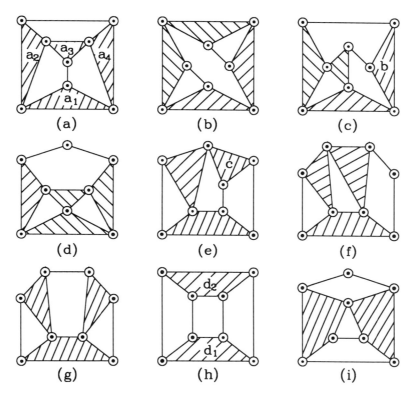

Figure 13.5 Atlas of (6,8) generalized chains of infinitely variable transmissions

2. For each case obtained in Step 1, identify the planet gear(s).
3. For each case obtained in Step 2, identify the corresponding carrier(s) for all possible cases.
4. For each case obtained in Step 3, identify the sun gear(s).
5. For each case obtained in Step 4, identify the gear pairs.
6. For each case obtained in Step 5, identify the revolute pairs.

13.5.1 Planetary gear trains with five members

For the two (5,6) generalized chains shown in Figure 13.4, only the one shown in Figure 13.4(a) is qualified to have a ground link, due to the constraints that a ground link should be a multiple link and should not be included in a three-bar loop. Its corresponding specialized planetary gear train can be identified as follows:
1. Since ternary links 1 and 3 are symmetric, based on the concept of similar classes, only one of them can be taken as the ground link.
2. If ternary link 1 is taken as the ground link, joints a, b, and c are revolute pairs and ternary link 3 is the planet gear.

3. Since binary links 2, 4, and 5 are symmetric and adjacent to the planet gear, based on the concept of similar classes, only one of them can be taken as the carrier.
4. If binary link 2 is taken as the carrier, joint d is revolute pair and binary links 4 and 5 are the sun gears.
5. Since there must be four revolute pairs and two gear pairs, the remaining two joints e and f are gear pairs.

Therefore, only one feasible specialized five-bar planetary gear train is available, as shown in Figure 13.6.

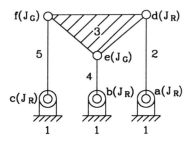

Figure 13.6 Feasible specialized five-bar planetary gear train

13.5.2 Planetary gear trains with six members

For the nine (6,8) generalized chains shown in Figure 13.5, feasible specialized chains are identified as follows:

Ground link

Since a ground link must be a multiple link that is not included in a three-bar loop in the chain, only generalized chains shown in Figures 13.5(a), (c), (e), and (h) are qualified to have the ground link. The ground link can be identified as follows:

1. For the generalized chain shown in Figure 13.5(a), based on the concept of similar classes, any one of ternary links a_1, a_2, a_3, and a_4 can be taken as the ground link. Figure 13.7(a) shows its corresponding specialized mechanism with link a_1 as the ground link. In this case, joints a, b, and c are revolute pairs.
2. For the generalized chain shown in Figure 13.5(c), only ternary link b can be taken as the ground link. Figure 13.7(b) shows its corresponding specialized mechanism. In this case, joints a, b, and c are revolute pairs.
3. For the generalized chain shown in Figure 13.5(e), only ternary link c can be taken as the ground link. Figure 13.7(c) shows its corresponding specialized mechanism. In this case, joints a, b, and c are revolute pairs.

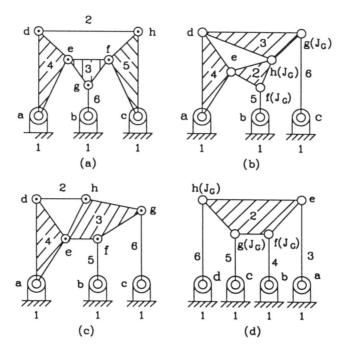

Figure 13.7 Feasible specialized six-bar planetary gear trains

4. For the generalized chain shown in Figure 13.5(h), based on the concept of similar classes, either quaternary link d_1 or quaternary link d_4 can be taken as the ground link. Figure 13.7(d) shows its corresponding specialized mechanism with link d_1 as the ground link. In this case, joints a, b, c, and d are revolute pairs.

Therefore, four specialized chains with one identified ground link are available as shown in Figures 13.7(a)-(d).

Planet gear

Based on the four specialized mechanisms with the ground link identified as shown in Figures 13.7(a)-(d), planet gears can be identified as follows:

1. For the one shown in Figure 13.7(a), since link 2 is not a multiple link, this case is out of consideration.
2. For the one shown in Figure 13.7(b), since ternary links 2 and 3 are not adjacent to the ground link and are multiple links, they are planet gears.
3. For the one shown in Figure 13.7(c), since link 2 is not a multiple link, this case is out of consideration.
4. For the one shown in Figure 13.7(d), since quaternary link 2 is not adjacent to the ground and is a multiple link, it is a planet gear.

Carrier

Since there must be a carrier corresponding to each planet gear and two planet gears in series must share a common carrier, carriers can be identified from the two specialized chains shown in Figures 13.7(b) and (d) as follows:
1. For the one shown in Figure 13.7(b), ternary link 4 is the common carrier to planet gears 2 and 3, and joints d and e are revolute pairs.
2. For the one shown in Figure 13.7(d), binary link 3 is the carrier to planet gear 2, and joint e is a revolute pair.

Sun gear

Since any link that is adjacent to the ground link and is not a carrier is a sun gear, sun gears can be identified from the two specialized chains shown in Figures 13.7(b) and (d) as follows:
1. For the one shown in Figure 13.7(b), binary links 5 and 6 are sun gears.
2. For the one shown in Figure 13.7(d), binary links 4, 5 and 6 are sun gears.

Revolute pairs

Since there must be five revolute pairs in each design, all revolute pairs for the two specialized chains shown in Figures 13.7(b) and (d) are already identified.

Gear pairs

Since there must be three gear pairs in each design, the remaining three unassigned joints f, g, and h in the two specialized chains shown in Figures 13.7(b) and (d) are gear pairs.

In conclusion, two feasible specialized chains are synthesized, as shown in Figures 13.7(a) and (b), that satisfy the design requirements and constraints.

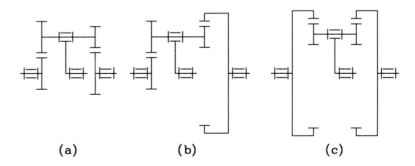

(a)　　　　　　　(b)　　　　　　　(c)

Figure 13.8 Atlas of five-bar planetary gear trains for infinitely variable transmissions

13.6 Particularization

The next step of the creative design methodology is to particularize each feasible specialized chain, by applying the generalizing rules backwards to obtain the corresponding schematic diagram of the planetary gear trains.

During the process of particularization, the two gears adjacent to a gear pair can be external or internal. By enumerating all possible variations of external and internal gears, numerous planetary gear trains can be obtained. For practical applications, it is very unlikely that a planet gear is in an internal form. With this additional constraint, Figure 13.8 shows the atlas of five-bar planetary gear trains for the feasible specialized planetary gear train shown in Figure 13.6, and Figure 13.9 shows the atlas of six-bar planetary gear trains for the two feasible specialized planetary gear trains shown in Figures 13.7(b) and (d).

13.7 Atlas of New Infinitely Variable Transmissions

The design configuration shown in Figure 13.8(b) is the original existing design. Therefore, with only external gears in the planets, the other ten concepts shown in Figure 13.8 and Figure 13.9 are new designs for the planetary gear trains, with five and six members, of infinitely variable transmissions.

13.8 Remarks

Although the creative design methodology presented in Chapter 6 provides a systematic step by step procedure for the configuration synthesis of mechanical devices, this methodology is quite flexible for practical applications. If many existing designs are available, design engineers can conclude design requirements and constraints by studying the existing designs. If existing designs are limited, design engineers can still conclude major design requirements and constraints based on available designs, and modify these requirements and constraints according to their engineering judgments. Furthermore, if no existing design is available, design engineers can even decide the design requirements and constraints on their own, possibly with the help of creative problem solving techniques as described in Chapter 5 or with the aid of the technique of quality function deployment for specification development.

If all the design concepts shown in Figure 13.8 and Figure 13.9 are not feasible in carrying out kinematic design and/or power flow analysis, generalized chains with seven links and ten joints, or even with higher numbers of links and joints, should be further considered for specialization and particularization.

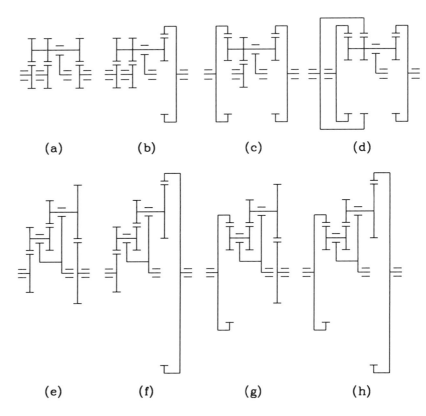

Figure 13.9 Atlas of six-bar planetary gear trains for infinitely variable transmissions

Problems

13.1 For each five-bar planetary gear train shown in Figure 13.8, identify its corresponding topology matrix.

13.2 For each six-bar planetary gear train shown in Figure 13.9, identify its corresponding generalized chain.

13.3 Carry out an exhaustive literature search for the applications of six-bar planetary gear trains with two degrees of freedom in automotive automatic transmissions.

13.4 Count the number of specialized chains, for six-bar planetary gear trains with identified ground link, for the nine (6,8) generalized chains shown in Figure 13.5.

13.5 Synthesize all possible design configurations that have the same topological characteristics as the planetary gear train shown in Figure 13.9(f).

13.6 Synthesize all possible design configurations that have the same topological characteristics as the planetary gear train shown Figure 13.1, if the requirement that the design should have seven members and the constraint that the ground link must have at least four incident joints are added.

13.7 Synthesize all possible design configurations that have the same topological characteristics as the planetary gear train shown Figure 13.1, if the requirement that the design should have eight members and the constraint that the ground link must have at least five incident joints are added.

13.8 Define design requirements and constraints for the conceptual design of the topological structures of planetary gear trains with two degrees of freedom for automotive automatic transmissions based on any creative technique provided in Chapter 5.

13.9 Synthesize all possible design configurations that have the required design requirements and constraints as concluded in Problem 13.8 for seven-bar planetary gear trains with two degrees of freedom for automotive automatic transmissions.

References

Buchsbaum, F. and Freudenstein, F., "Synthesis of kinematic structure of geared kinematic chains and other mechanisms," *Journal of Mechanisms*, Vol. 5, 1970, pp. 357-392.

Harary, F., Graph Theory, Addison-Wesley, 1969.

Yan, H. S., "A methodology for creative mechanism design," *Mechanism and Machine Theory*, Vol. 27, No. 3, 1992, pp. 235-242.

Yan, H. S. and Hsieh, L. C., "Concept design of planetary gear trains for infinitely variable transmissions," Proceedings of 1989 International Conference on Engineering Design, Harrogate, United Kingdom, August 22-25, 1989, pp. 757-766.

CHAPTER 14

CONFIGURATIONS OF MACHINING CENTERS

A *machining center* is a machine tool consisting of four basic components: a spindle, a tool magazine, a tool change mechanism, and a machine tool structure including motion transmissions of various axes. The machine tool structure largely determines its machined surface, stiffness, and dynamic quality. The spindle holds and rotates the tool to machine the workpiece to the desired surface. The tool magazine stores the tools and moves them to suitable positions for uses in tool exchanging operations. The tool change mechanism executes tool changes between the tool magazine and the spindle. It generally comprises of a tool transfer arm (T_a), a parking station (P_s), and a tool change arm (T_c) for performing tool transportation, tool rotary, and tool exchange motion, respectively.

The system that automatically performs tool changes between the spindle and the tool magazine of a machining center is called *automatic tool changer* (ATC). The simplest automatic tool changer is the design with a tool magazine parallel to the spindle and without a tool change arm (even without a parking station and a tool transfer arm), and the relative motion between the tool magazine and the spindle achieves tool change motions. Figures 14.1(a) and (b) show two 3-axis horizontal machining centers without tool change arms, and Figure 14.1(c) shows a 3-axis horizontal machining center with a tool change arm.

This chapter generates all possible configurations of machining centers without tool change arms subject to topology and motion requirements. The creative design methodology presented in Chapter 6 is modified, as shown in Figure 14.2, for this purpose.

212 ■ 14. Configurations of Machining Centers

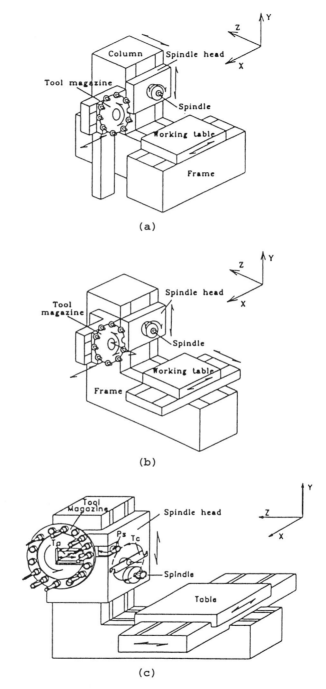

Figure 14.1 Three-axis horizontal machining centers

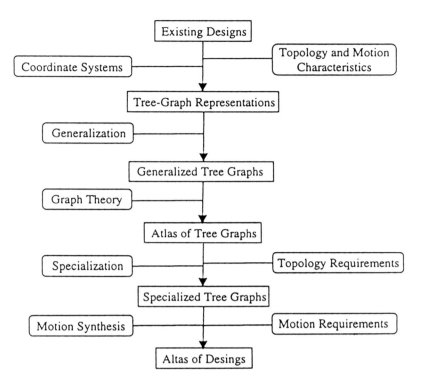

Figure 14.2 Creative design methodology for the configurations of machining centers

14.1 Existing Designs

The first step of this creative design methodology is to study available existing designs, to conclude their topology and motion characteristics.

Since the main function of a tool change mechanism is to perform tool change motions successfully, it is important to analyze the relations between the topological structure and the tool change motions. The tool change mechanism shown in Figure 14.1(a) is taken as an example. Its sequence of tool change motion is as follows:
1. The tool magazine moves along the negative direction of the X axis to grasp the old tool in the spindle.
2. The column moves along the positive direction of the Z axis to leave the old tool.
3. The tool magazine rotates about the Z axis to make the new tool in front of the spindle.

4. The column moves along the negative direction of the Z axis to insert the new tool into the spindle.
5. The tool magazine moves along the positive direction of the X axis to leave the new tool away.

The tool change motions of the machining center shown in Figure 14.1(b) are similar to that shown in Figure 14.1(a), except that it uses the tool magazine to extract and insert the tools. The tool change motions of these two machining centers are shown schematically as Figures 14.3(a) and (b), where M and S represent the tool magazine and the spindle, respectively. For the sake of simplicity here, P, R and C represent the prismatic, revolute and cylindrical pairs, respectively. And the number inside a circle means the sequence of tool change motions.

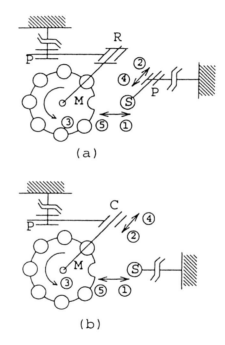

Figure 14.3 Tool change motions

By information search, analyzing available existing designs, literature and patents, and even consulting experts, the topology and motion characteristics of 3-axis horizontal machining centers with tool magazines parallel to the spindle and without a tool change arm are concluded as follows:

Topology characteristics
1. They are spatial open-type mechanisms, i.e., the corresponding chains are not closed, with multiple degrees of freedom.
2. They have one ground link (frame).
3. They have a spindle that is an end member.
4. They have a working table, where the number of joints between the spindle and the working table is four.
5. They have a tool magazine that is an end member extended from the member that is located from the frame to the spindle head.
6. The end members could be the spindle, the tool magazine, or the working table. And the maximum number of end members must be three.
7. The joint incident to the spindle must be a revolute pair.
8. The joints between the spindle head and the working table must be prismatic pairs.
9. The joints between the tool magazine and the extending member are revolute, prismatic, or cylindrical pairs. And there must be a revolute pair or a cylindrical pair incident to the tool magazine.

Motion characteristics
1. The working table has three relative motions in X, Z, and Y directions, sequentially with respect to the spindle head.
2. The automatic tool changer uses the relative motions between the tool magazine and the spindle to exchange tools. The tool change motion sequence is $P_x \to P_z \to R_z \to P_z \to P_x$, where P and R represent the sliding motion and rotary motion, respectively, and subscripts x, y, and z represent the directions of these motions.
3. In order to achieve tool change motions, the relative degrees of freedom between the tool magazine and the spindle head must be at least three.

14.2 Tree-graph Representations

The second step of the creative design methodology is to represent the existing designs by their corresponding tree-graphs.

In order to represent the topological structures of machining centers, a coordinate system is defined to describe the allocation of each motion axis of machining centers, based on the International Organization for Standardization (ISO). This standard coordinate system is a right-handed rectangular Cartesian one, related to a work piece mounted in a machine and aligned with the principal linear sideways of that machine. The positive direction of movement of a component in the machine is that which causes an increasing positive dimension of the work piece.

The schematic drawings of horizontal machining centers appended to ISO standard are shown in Figure 14.1.

Based on defined coordinate systems, the mechanism can be described by representing its links and joints with nodes and edges, respectively, in which two nodes (edges) are adjacent whenever the corresponding links (joints) of the mechanism are adjacent.

According to this representation, the links of a machining center are represented by nodes, with the names of links as shown in Table 14.1, and the joints of a machining center are represented by edges with the name of the type of joints, as shown in Table 14.2. The name of joints has a subscript that denotes the allocation of motion axis of that joint. If the motion axis of a revolute joint is parallel to the X axis, this joint is denoted as R_x; if the motion axis of a revolute joint is parallel to the Y axis, this joint is denoted as R_y, and so on. Figures 14.4(a) and (b) show the corresponding tree-graph representations of the machining centers shown in Figures 14.1(a) and (b), respectively.

Table 14.1 Graph representations of links

Links	Symbols
Frame (Ground link)	⊙F_r
Spindle	• S
Working table	• T
Tool magazine	• M
Connecting rod	• L_1 • L_2 ...

Table 14.2 Graph representations of joints

Joints	Degrees of freedom	Symbols
Revolute joint	1	R_x R_y R_z ...
Prismatic joint	1	P_x P_y P_z ...
Cylindrical joint	2	C_x C_y C_z ...

14.3 Generalized Tree-Graphs

The third step of the creative design methodology is generalization. Its purpose is to transform the original mechanism, that involves various types of members (nodes) and joints (edges), into a generalized tree-graph. The process of generalization is based on a set of generalizing rules. These generalizing rules are derived according to defined generalizing principles. The generalizing principles and rules are described in detail in Chapter 7.

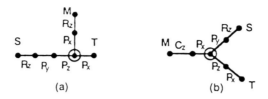

Figure 14.4 Tree-graph representations of machining centers

For the tree-graphs of machining centers shown in Figure 14.4, the generalization is carried out by using the following steps:
1. The frame is released and generalized into a ternary link, i.e., a node with three incident edges.
2. The end members, such as the spindle, the tool magazine, and the working table, are all generalized into singular links, i.e., nodes with only one incident edge.
3. The other members are generalized into binary links, i.e., nodes with two incident edges.
4. The prismatic pairs are generalized into revolute pairs.

For the two designs shown in Figures 14.4(a) and (b), their corresponding generalized tree-graphs, with seven nodes and six edges, are shown in Figures 14.5(a) and (b), respectively.

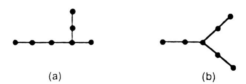

Figure 14.5 Generalized tree graphs of machining centers

14.4 Atlas of Tree-Graphs

The fourth step of the creative design methodology is to generate all possible tree-graphs with the given numbers of edges and vertices. For an open loop mechanism, its number of joints is one fewer than its number of links. According to graph theory, the explicit numbers of tree-graphs through p nodes can be obtained from the coefficients of the following polynomial:

$$T(x) = x + x^2 + x^3 + 2x^4 + 3x^5 + 6x^6 + 11x^7 + 23x^8 + 47x^9 + 106x^{10} + 235x^{11} + 551x^{12} + \ldots \quad (14.1)$$

Equation (14.1) indicates that the numbers of tree-graphs with 1, 2, 3, 4, 5, 6, 7, 8, 9, 10, 11, 12 nodes are 1, 1, 1, 2, 3, 6, 11, 23, 47, 106, 235, 551, respectively. Figure 14.6 shows the atlas of tree-graphs with $p=2$ to $p=7$.

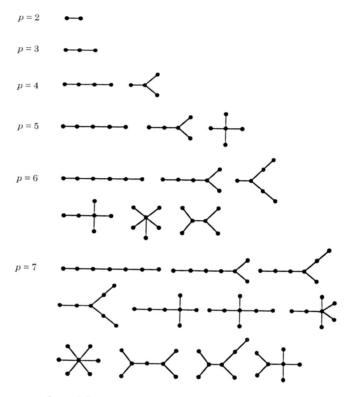

Figure 14.6 Atlas of tree graphs with 2 to 7 nodes

14.5 Specialized Tree-Graphs

The fifth step of the creative design methodology is specialization. Its purpose is to assign specific types of members and joints into available atlas of tree-graphs to generate all possible non-isomorphic specialized tree-graphs, subject to topology constraints. Chapter 10 describes the process of specialization in detail. Topology requirements of members and joints of machining centers in their corresponding tree graphs are listed first. Then, all possible topology structures can be obtained after specialization.

Topology requirements

Topology requirements are concluded according to the topology characteristics of existing designs. Here, the design requirements of members and joints of 3-axis horizontal machining centers in their corresponding tree-graphs are:
1. For a 3-axis machining center, its corresponding tree-graph must have at least six nodes.
2. The maximum number of pendant nodes must be three.
3. There must be a pendant node as the spindle.
4. There must be a node, where the path length to the spindle is four, as the working table.
5. There must be a rooted node, located on the path from the spindle head to the working table, as the frame.
6. There must be a node, located on the pendant node extending from the node that is located from the frame to the spindle head, as the tool magazine.
7. The pendant nodes must be the spindle, the tool magazine, or the working table.
8. The edge incident to the spindle must be assigned as a revolute pair.
9. The edges between the spindle head and the working table must be assigned as prismatic pairs.
10. The edges between the tool magazine and the branch node must be assigned as revolute, prismatic, or cylindrical pairs. There must be a revolute pair or a cylindrical pair incident to the tool magazine.

Specialization

For the tree-graphs shown in Figure 14.6, only the graphs with at least six nodes and at most three pendant nodes satisfy the topology requirements. These feasible tree-graphs are shown in Figure 14.7. In the following, the spindle, the working table, the frame, and the tool magazine are assigned into those tree-graphs shown in Figure 14.7 up to six links. Then, specific types of joints are assigned into the tree-graphs according to topology requirements.

a. Spindle

For the tree-graph shown in Figure 14.7(a), only the pendant node can be assigned as the spindle. Hence, the assignments of the spindle (S) produces only one non-isomorphic result, as shown in Figure 14.8(a). Similarly, for the tree graphs shown in Figures 14.7(b) and (c), the assignment of the spindle produces four non-isomorphic results, as shown in Figures 14.8(b)-(e).

b. Working table

For the tree-graph shown in Figure 14.8(a), only the node where the path length to the spindle is four can be assigned as the working table (T). Hence, the assignments of the working table produces only one result, as shown in Figure 14.9(a). Similarly, the tree-graphs shown in Figures 14.8(b), (c), and (e) produce three results, as shown in Figures 14.9(b), (c), and (d), respectively. The tree-graph shown in Figure 14.8(d) has no working table available, since no node where the path length to the spindle is four.

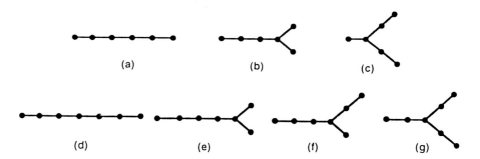

Figure 14.7 Feasible atlas of tree-graphs of machining centers

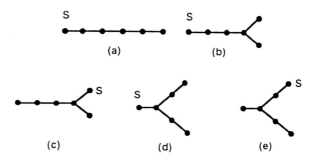

Figure 14.8 Assignments of the spindle

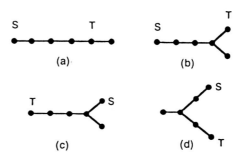

Figure 14.9 Assignments of the working table

c. Frame

For the tree-graph shown in Figure 14.9(a), only the node that is located on the path from the spindle head to the working table can be assigned as the frame (Fr). Hence, the assignments of the frame produces four results, as shown in Figures 14.10(a)-(d). Similarly, for the tree-graphs shown in Figures 14.9(b)-(d), the assignments of the frame produces twelve results, as shown in Figures 14.10 (e)-(p).

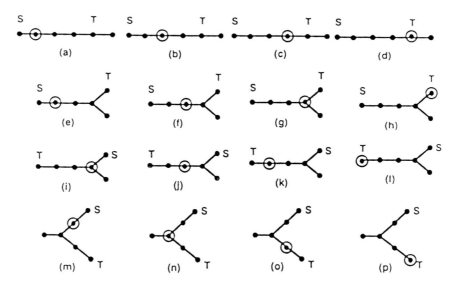

Figure 14.10 Assignments of the frame

d. Tool magazine

For the tree-graphs shown in Figure 14.10, only the node that is located on the pendant node extending from the node that is located from the frame to the spindle head can be assigned as the tool magazine. For the tree-graphs shown in Figures 14.10(d), (g), (h), (i), (j), (k), (l), (n), (o), and (p), the assignments of the tool magazine produce ten results, as shown in Figures 14.11(a)-(j), respectively. The tree-graphs shown in Figures 14.10(a), (b), (c), (e), (f), and (m) do not satisfy the topology requirements.

e. Joint specialization

Based on the topology requirements, the edge incident to the spindle and the edges between the spindle head and the working table must be assigned as a revolute pair and a prismatic pair, respectively. Then, the joint permutation can be assigned to the edges between the tool magazine and the branch vertex. According to the topology requirements and the joint permutations, the tree-graphs shown in Figure 14.11 can be specialized, and all feasible topology structures of 3-axis machining centers are obtained as shown in Figure 14.12.

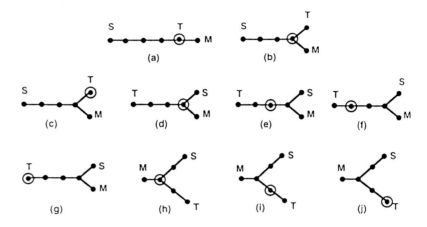

Figure 14.11 Assignments of the tool magazine

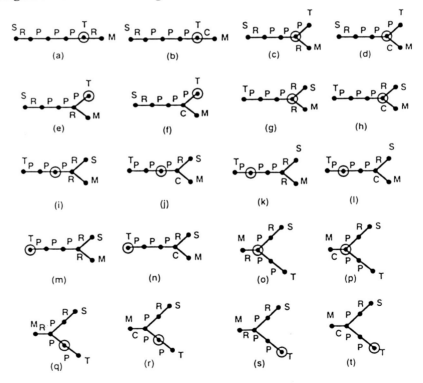

Figure 14.12 Assignments of the joints

14.6 Atlas of Machining Centers

The last step of this creative design methodology is motion synthesis. Its purpose is to assign axis directions of joints into available specialized tree-graphs to generate feasible mechanisms subject to certain motion requirements.

Motion requirements are concluded based on the motion characteristics of the designs. The motion requirements of the mechanism of 3-axis horizontal machining centers are:

1. In order to achieve the tool change motions, the relative degrees of freedom between the spindle head and the tool magazine must be at least three.
2. The direction of the spindle is defined as the Z axis by the coordinate systems of the machine tool.
3. The motion axes from the spindle head to the working table are assigned as Y, Z, and X axes, respectively.
4. The tool change motions shown in Figure 14.3 can be concluded and the tool change motion sequence is $P_x \to P_z \to R_z \to P_z \to P_x$.

Therefore, the first step of motion synthesis is to check the relative degrees of freedom between the tool magazine and the spindle head, then delete specialized tree-graphs that do not satisfy these design requirements. The second step is to assign the directions of motion axes of the machine tool structure to the specialized tree-graph. The third step is to allocate the directions of tool change motion axes according to tool change motion sequence.

According to the motion requirements, the relative degrees of freedom between the tool magazine and the spindle head are at least three, to achieve tool change motions. For the tree-graphs shown in Figures 14.12(g), (h), (i), (j), (k), (l), (m), (n), (o), (q), and (s), the relative degrees of freedom between the spindle head and the tool magazine are less than three. Therefore, only Figures 14.12(a), (b), (c), (d), (e), (f), (p), (r), and (t) are feasible topological structures that satisfy the design requirements of tool change motions.

Furthermore, without considering the direction of the displacement, the tool change motion sequence is symmetrical. Therefore, for the motion synthesis of tool change mechanisms, only half of the tool change motion that can be separated into three parts are necessary to be considered: grasping tool, extracting tool, and changing tool. Based on the study of tool change motion sequence, the following points are concluded:

1. For the motion of grasping tool, there must be a relative sliding degree of freedom in the X direction between the spindle head and the tool magazine.

2. For the motion of extracting tool, there must be a relative sliding degree of freedom in the Z direction between the spindle head and the tool magazine.
3. For the motion of changing tools, there must be a relative rotary degree of freedom in the Z direction between the spindle head and the tool magazine.

Therefore, the procedure for the tool change motion synthesis are summarized as follows:

Step 1. If there is no relative degrees of freedom between the branch node and the tool magazine that can be assigned as the tool change motions, go to Step 4; otherwise continue.

Step 2. If there is no relative degrees of freedom between the spindle head and the branch node that can be assigned as the axis of tool change motions, assign this relative degrees of freedom between the branch node and the tool magazine as the tool change motions, go to Step 5; otherwise continue.

Step 3. If there is a relative degrees of freedom between the spindle head and the branch node, assign this relative degrees of freedom between the branch node and the tool magazine or between the branch node and the spindle head as the tool change motions and go to Step 5.

Step 4. If there is a relative degrees of freedom between the branch node and the spindle head that can be assigned as the tool change motions, go to Step 5; otherwise delete this specialized tree-graph.

Step 5. Continue to complete the motion synthesis, go to Step 1; otherwise stop.

If there are redundant degrees of freedom between the branch node and the tool magazine that is not assigned as the tool change motions, delete this specialized tree-graph.

In what follows, the motion synthesis subject to the design requirements of tool change motions is carried out:

1. For the specialized tree-graphs shown in Figures 14.12(a), (b), (c), (d), (e), (f), (p), (r), and (t), the assignments of the spindle axis and the machine tool axes parallel to axes Z, Y, Z, and X, respectively, are shown in Figure 14.13.
2. For the tree-graph shown in Figure 14.13(a), there is no relative degree of freedom between the branch node and the tool magazine that can be assigned as the grasping tool motion; go to Step 4.
3. There is a relative degree of freedom between the spindle head and the branch node in the same direction of grasping tool motion that can be assigned as tool change motions; go to Step 5.
4. Continue the tool change motion synthesis and go to Step 1.

5. For the tree-graphs shown in Figure 14.13(a), there is no relative degree of freedom between the branch node and the tool magazine that can be assigned as the extracting tool motion; go to Step 4.
6. There is a relative degree of freedom between the spindle head and the branch node in the same direction of extracting tool motion that can be assigned as tool change motions; go to Step 5.
7. Continue the tool change motion synthesis and go to Step 1.
8. For the tree-graph shown in Figure 14.13(a), there is a relative degree of freedom between the branch node and the tool magazine that can be assigned as the changing tool motion; go to Step 2.

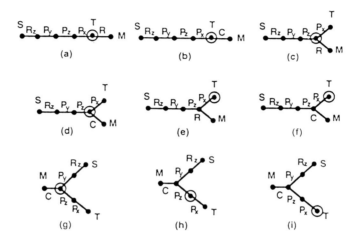

Figure 14.13 Allocation of machine tool axes

9. Since there is no relative degree of freedom between the spindle head and the branch node in the same direction of changing tool motion that can be assigned as tool change motions, assign this relative degree of freedom between the branch node and the tool magazine as the tool change motions; go to Step 5.
10. Complete the motion synthesis, and the topological structures of machining centers are obtained as shown in Figure 14.14(a).

Figure 14.14(a) has no redundant degree of freedom between the branch node and the tool magazine. Hence, it satisfies the design requirements of tool change motions.

For motion synthesis of machining centers, the edges of tree-graphs are assigned as the type of actuated joints for tool change motions. According to tool change motion sequence, feasible specialized tree-graphs are assigned to their corresponding mechanisms step by step. Some topological structures that cannot satisfy the tool change motions are removed.

226 ▪ 14. Configurations of Machining Centers

Figure 14.14 Tree-graphs of machining centers with six links

For the specialized tree-graphs shown in Figure 14.13, the tool change motion synthesis produces two results, as shown in Figure 14.14. Similarly, the mechanisms of machining centers with seven links could be synthesized by the same approach. For mechanisms of machining centers with seven links, thirteen types are available, as shown in Figure 14.15. The schematic drawings of machining centers with six and seven links are shown in Figure 14.16 and Figure 14.17, respectively.

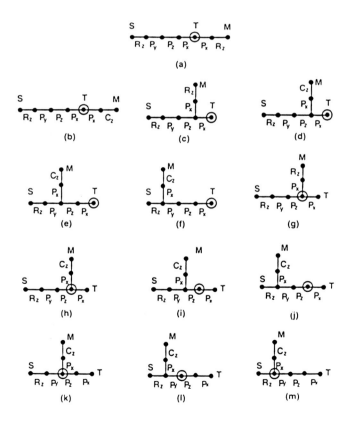

Figure 14.15 Tree-graphs of machining centers with seven links

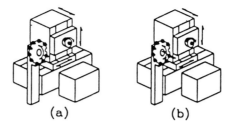

Figure 14.16 Configurations of machining centers with six links

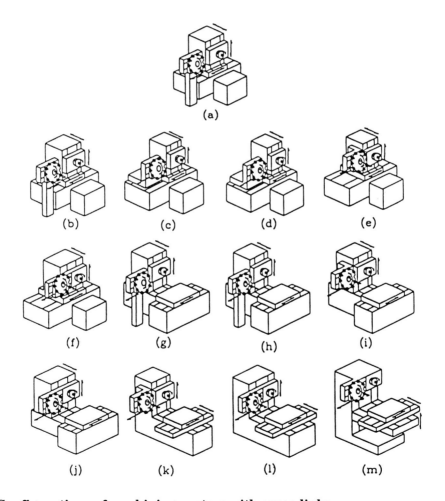

Figure 14.17 Configurations of machining centers with seven links

14.7 Remarks

The creative design methodology presented in Chapter 6 is a core methodology for the generation of all possible design concepts with desired topological characteristics. This methodology can be further modified and/or expanded, based on designers' needs for the considerations of other requirements and constraints, such as motion characteristics presented in this chapter.

Problems

14.1 Search literature regarding automatic tool changers of machining centers.
14.2 Search patents of automatic tool changers with tool change arms for machining centers.
14.3 Locate an existing automatic tool changer with a tool change arm and analyze its performance mathematically.
14.4 Provide checklist questions for the improvement of performance of the existing automatic tool changer in Problem 14.3.
14.5 List major attributes of automatic tool changers with tool change arms based on knowledge gained from Problems 14.1-14.4.
14.6 Conduct a group brainstorming session to conclude the design requirements and constraints of automatic tool changers with tool change arms based on knowledge gained from Problems 14.1-14.5.
14.7 Count the numbers of specialized chains with identified spindle, for automatic tool changers without a tool change arm, from the atlas of tree-graphs shown in Figure 14.7.
14.8 Synthesize all possible design configurations with eight links that have the same topological and motion characteristics as the three-axis horizontal machining centers shown in Figure 14.1.
14.9 Based on Problems 14.1-14.6, synthesize design configurations of machining centers with tool change arms.

References

Chen, F. F., Mechanism Configuration Synthesis of Machining Centers, Ph.D. dissertation, Department of Mechanical Engineering, National Cheng Kung University, Tainan, Taiwan, January 1997.

Den, N., Graph Theory with Application to Engineering and Computer Science, Prentice-Hall, 1974.

Harary, F., Graph Theory, Addison-Wesley, 1969.

ISO, Numerical Control Machines - Axis and Motion Nomenclature, ISO 841, International Organization for Standardization, Switzerland.

Shinno, H. and Ito, Y., "Generating method for structural configuration of machine tools," *JSME Transactions (C)*, Vol. 50, No. 449, 1983, pp. 213-221.

Shinno, H. and Ito, Y., "Computer aided concept design for structural configuration of machine tools: variant design using directed graph," ASME Transactions, *Journal of Mechanisms, Transmissions, and Automation in Design*, Vol. 109, 1987, pp. 372-376.

Yan, H. S., "A methodology for creative mechanism design," *Mechanism and Machine Theory*, Vol. 27, No. 3, 1992, pp. 235-242.

Yan, H. S. and F. C. Chen, "Configuration synthesis of machining centers without tool change arms," *Mechanism and Machine Theory*, Vol. 33, No. 1/2, 1998, pp. 197-212.

APPENDIX

Figure 1.3 Machining center – an engineering design (Courtesy of Leadwell Company)

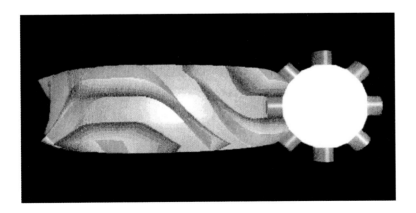

Figure 1.13 A roller gear cam

232 ▪ Appendix

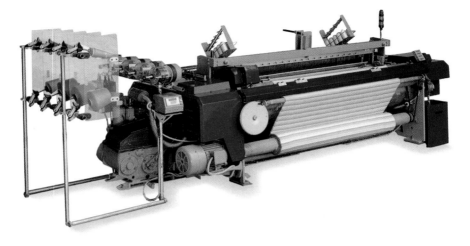

Figure 4.1 An existing rapier type shuttleless loom (Courtesy of I-Chin Company)

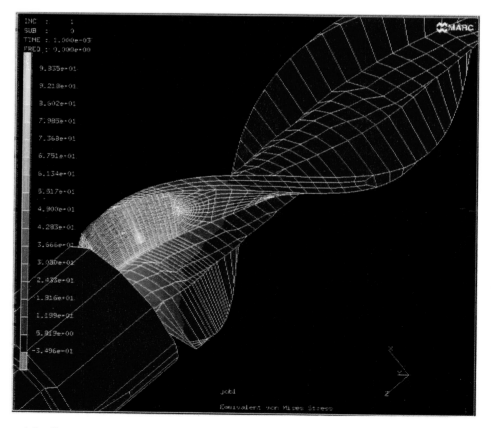

Figure 4.3 Contact stress of the variable pitch screw

Appendix • 233

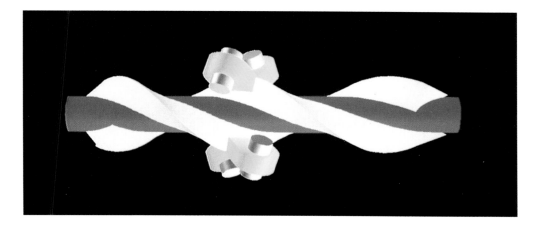

Figure 4.4 Geometric solid modeling of the variable pitch screw

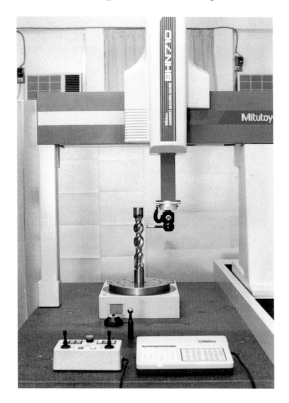

Figure 4.5 Precision measurements of the variable pitch screw

Figure 4.6 Dynamic testing of the variable pitch screw transmission mechanism

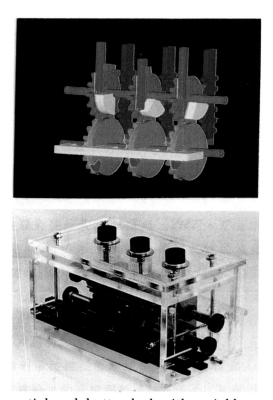

Figure 5.3 A new sequential push button lock with variable passwords

Appendix ▪ 235

Figure 5.6　A walking mechanical horse

INDEX

a

actuator	20
adapt	57
analysis	4
analysis of existing designs	49
applied force	21
attribute	67, 147
attribute listing	67
automatic tool changer	211
automorphic	140

b

basic rigid chain	136
belt	20
binary link	18
block	113
brainstorming	73
bridge-link	23
bridge-node	123

c

cam	20
cam pair	21
chain (link)	23
chain (sprocket)	20
chain-group	141
checklist	56
checklist method	56
checklist question	56
checklist transformation	57
clamping device	173
closed chain	23
closed (graph)	123
combine	57
compatibility constraint	147
concept	9

conceptual design	9
connected (chain)	23
connected (graph)	123
constrained motion	25
contracted link	138
contracted link adjacency matrix	138
contracted link assortment	145
creation	9
creativity's characteristics	41
creative enhancement	45
creative process	36
creativity	35
cultural barrier	42
cycle	140
cycle index	165
cycle structure representation of a permutation	165
cylinder	20
cylindric pair	22

d

degenerate kinematic chain	136
degrees of constraint	25, 27
degrees of freedom	25, 27
design	3
design process	8
direct contact	23
disconnected (chain)	23

e

element	21
emotional barrier	42
engineering	3
engineering design	3
execution phase	38
experimental tests and measurements	52

f

feasible specialized chain	159
file of experts	55
flat pair	23
follower	20

g

gear	19
gear pair	21
generalization	91, 97
generalized chain	106, 117
generalized joint	97
generalized kinematic chain	106
generalized link	99
generalized mechanical device	106
generalized revolute joint	102
generalized tree graph	216
generalizing principle	99
generalizing rule	99
graph	123
ground link	23

h

helical pair	22

i

illumination phase	38
incidence joint sequence	146
incubation phase	37
infinitely variable transmission	197
information search	52
innovation	35
invention	35
inventory	166
isomorphic	28
isomorphism	28

j

joint	21
joint element	138
joint-group	141

k

kinematic chain	23, 117, 135
kinematic link	18
kinematic matrix	137
kinematic pair	21

l

labelled chain	138
labelled link adjacency matrix	138
line graph	124
link	18
link adjacency matrix	137
link assortment	120
link-chain	23
link element	138
link-group	140
literature search	52

m

machine design	4
machining center	211
magnify	58
mathematical analysis	50
matrix technique	83
mechanical design	4
mechanical device	17
mechanical engineering design	4
mechanical member	17
mechanism	17, 24, 25
mechanism design	4
methodology	87
minify	59
modify	60
morphological analysis	69
morphological chart analysis	69
morphology	69
motorcross suspension mechanism	185
multiple generalized joint	97

n

number synthesis	92

o

open chain	23

p

particularization	94
patent search	54
path	23

Index ■ 241

perceptual barrier 43
permutation 140
permutation group 140, 141
piston 20
planar block 124
planetary gear train 197
power screw 20
preparation phase 36
prismatic pair 21
pulley 20
put to other uses 62

q

quaternary link 18

r

rearrange 60
reverse 60
revolute pair 21
rigid chain 24, 117, 136
roller 19
rolling pair 21

s

screw pair 22
separated link 17
shock absorber 20
similar 141
similar class 141
simple generalized joint 97
simple kinematic chain 135
singular link 17
slider 17
sliding pair 21
specialization 92, 159
specialized chain 159
specialized tree graph 218
specification 89
spherical pair 22
spring 20
sprocket 20
structure 17, 24, 25
substitute 61
synthesis 5

t

ternary link	18
topological structure	17, 28
topology matrix	28
tree-graph representation	215
turning pair	21

u

universal joint	23

w

walk	23
wrapping joint	21